The Active Brain

By Nelunika Gunawardena Rajapakse

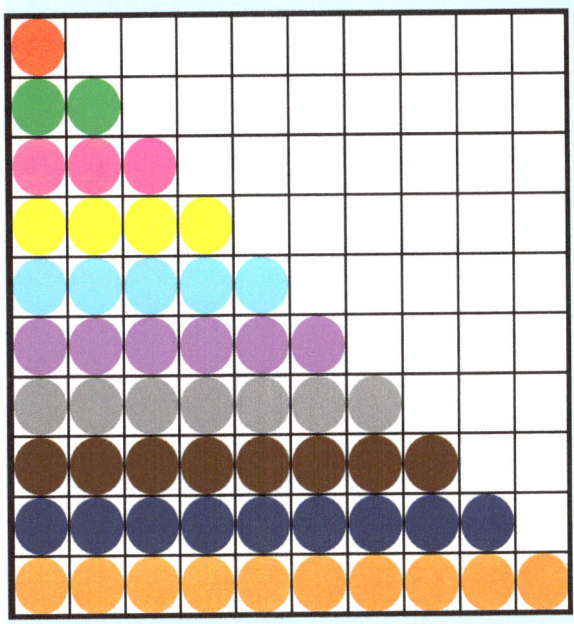

A Vision for
Intellectual
Wellness

Book 1

The Basic Number
Concepts from 1-9

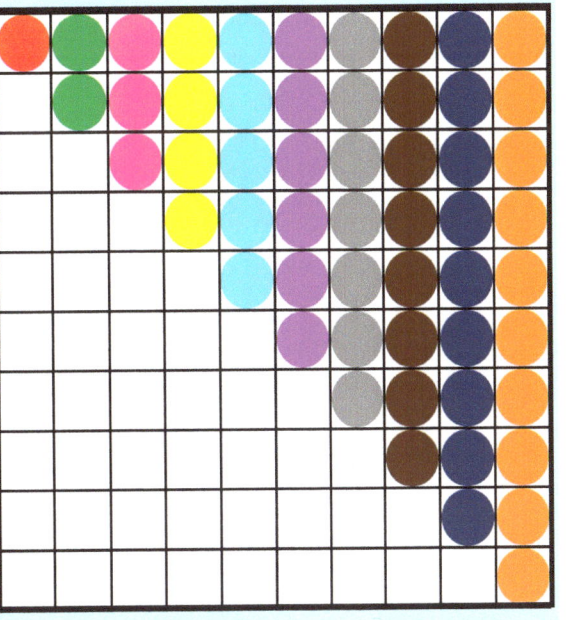

Precious Jewels of Mrs. "G"
Engaged Partnerships in Learning

The Active Brain

A Vision for Intellectual Wellness

Book 1

The Basic Number Concepts from 1-9

Precious Jewels of Mrs. "G"
Engaged Partnerships in Learning

By Nelunika Gunawardena Rajapakse

A Vision for Intellectual Wellness

Its Cultivation and Preservation

For All Ages

This is a precious gift of LOVE to humanity
In tribute to my beloved parents,
Dionicious and Mildred Gunawardena

Table of Contents

Author's Foreword

The Active Brain (Book 1) is a mathematical exposition from 'The Precious Jewels of Mrs. "G"' collection of offerings that have been thoughtfully created for the benefit of people of all ages, beginning with the universal child—not forgetting their mentoring counterparts dedicated to intellectual wellness and education.

It is a graphic presentation of mathematical wisdom, color-defined to bring forth precise concepts, their patterns, and relationships.

This edition, in particular, focuses upon the very first step along the hierarchical sequence of mathematical evolution—**The Number Concepts.**

The Active Brain is a series of volumes that seeks to systematically present precise concepts, and their ever-evolving patterns and relationships in order to keep the brain-mind stimulated along its encounters of exploration, discovery, and interactivity with the universe.

The world of relationships and relativity begins to evolve from the very inception of life. Our early childhood sensory experiences associated with practical life-impressions, no doubt, initiate in setting a concrete foundation towards the cultivated recognition of all things and events—as phenomena, in existence. Underlying these, indeed, are the secrets to discovery—by way of concepts, patterns, and relationships—that signify our world of interdependence.

Hence, it is important that we continually strive to remain centered—through each phase in our lives—so that we may sustain an acute vision and understanding of our dynamic world of phenomena. However, shared knowledge in partnership with selflessly mentoring counterparts and their unrelenting guidance is pivotal to the process of learning.

How best may we relate to change while we persist with the effort to consciously and conscientiously interact with our dynamic world?

The brain-mind awaits stimulation incessantly. It must remain sparked to direct and execute its various vital functions with respect to the body, and its environment.

Accordingly, **The Active Brain** is <u>a series</u> of enlightening and stimulating presentations focused upon the cultivation, advancement, and preservation of the intellect.

The mathematical mind is our natural human endowment; it must be nurtured with awareness and diligence—especially on the part of all mentors who impart knowledge.

Furthermore, it behooves us to recognize that our human consciousness is an evolving phenomenon that is interactively engaged with the universe, through infinite time. The universe, therefore, must be the very basis upon which human understanding and imagination must rest—with precision and exactitude—in regard to the order, chaos, and probabilities of its spatial relationships.

Consequently, our journey of precision in exploration must find its roots, at a concrete level—catering to the tender-most, curious mind blessed with a unique capacity to absorb and assimilate the physical attributes of our universe. The seeds of knowledge sown in such a fertile field of absorption will only mature with zest, in their own time, provided that we, 'the gardeners' continue to water and nourish the evolving seedlings.

Our universe, however, is to be analyzed, at a later stage of human development, in terms of deduction through analyses of more abstract properties, conceptual patterns, and their complex relationships in relation to the magnanimous wonders unceasingly manifested as phenomena in existence.

The color-defined exercises in this edition—***The Active Brain*** (<u>Book 1</u>)—are intended to keep the brain-mind engagingly stimulated by means of hand-eye coordination.

The sharers of information, and their participants at the receiving end, are to be engaged, in collaboration, since knowledge must be imparted as a modeled life-experience as opposed to being a lesson of teaching.

A substantial benefit from these presentations is possible, if and when the exercises are performed patiently and willingly to suit each individual's needs. Consistent practice, indeed, by means of these exercises of repetition—<u>presented each time, with a new twist</u>—as shown herein, will prepare you to further absorb and assimilate **the dynamics of the basic number concepts 1-9** indispensable for the cultivation of deeper understanding in regard to the many tiers of intellectual mathematical evolution.

This book, no doubt, is a thoughtful <u>first</u> gift towards the establishment of 'lovingly engaged partnerships in learning'.

It is recommended that its application be individual-centered in order to derive its fullest effectiveness.

Activity—that engages all five of the senses—sows lasting imprints in every intellect. All in all, we tend to absorb knowledge almost effortlessly through impressionistic life-experiences while we journey through the many phases of our lives.

Therefore, it must be emphasized that this book is intended primarily to involve prehensile movement—as hand-eye coordinated, intellectually stimulating activity so that we may expediently reach out to all ages.

Considered overall, engagement in collaboration, as a modeled life-experience—quite contrary to being a lesson of teaching—is the flame that kindles and sustains the light of knowledge to enliven the intellect, eventually to keep our hard-wired mind, the human brain, actively sharp.

Beneficial Instructions for Carrying out the Exercises in this Book

Necessary Implements

1. A set of well-sharpened colored pencils (one of each) in the primary colors (red, blue, yellow), secondary colors (green, purple, orange), and tertiary colors (brown, pink, gray), also to include black and white

2. Two #2 pencils

3. A handmade pencil holder of heavy grade paper (Fold accordion style to create 15 zigzag grooves.)

Implementation

1. Help the learner internalize the quantities defined by color; it is to associate the color with the quantity.

2. Coloring—to fill in the circles and squares in the exercises as shown in this book—must take place in <u>slow and gentle</u> downward to upward strokes.

3. The process of coloring must convey a silent voice;

    "Each line must touch the preceding line."

    "We color **lightly, slowly, gently, and evenly** from the top line to the bottom line and from the bottom line to the top line."

    "We always color **patiently and precisely** from the top line to the bottom line, and from the bottom line to the top line."

"We must focus our attention on the pencil point alone so that the pencil strokes are even."

"We must never rush to fill in the circles or squares."

"We must make every effort to keep the strokes within the circumference of each circle and the perimeter of each square that we fill in."

"We must take pride in seeing the outcome of our efforts of patience."

Section I

Cardinal Numbers in Vertical Order

Presentation Chart 1

<u>Cardinal</u> Numbers in Vertical Order

Illustrated below are the <u>quantities</u> and associated <u>numerical symbols</u> of <u>cardinal numbers</u>. (Note the vocabulary underlined.)

- Observe the quantities arranged along a vertical sequence.
- Observe the numerals associated with the quantities.
- Observe the particular color assigned to each quantity.

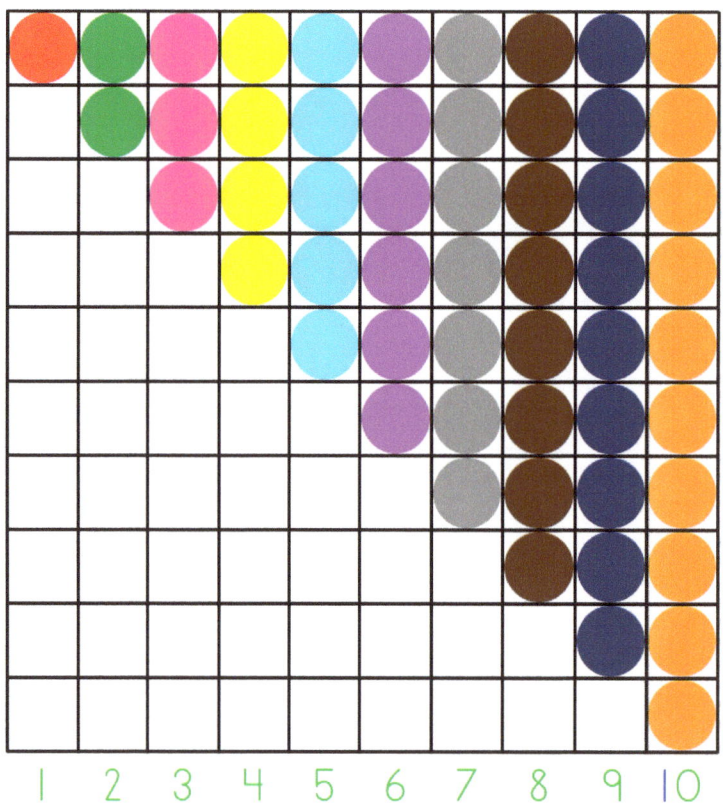

A Vision for Intellectual Wellness
In tribute to Dionicious and Mildred Gunawardena

Exercise 1-1

Fill in the open spaces with the missing colored bead bars (or <u>quantities</u>) in their vertical arrangement.

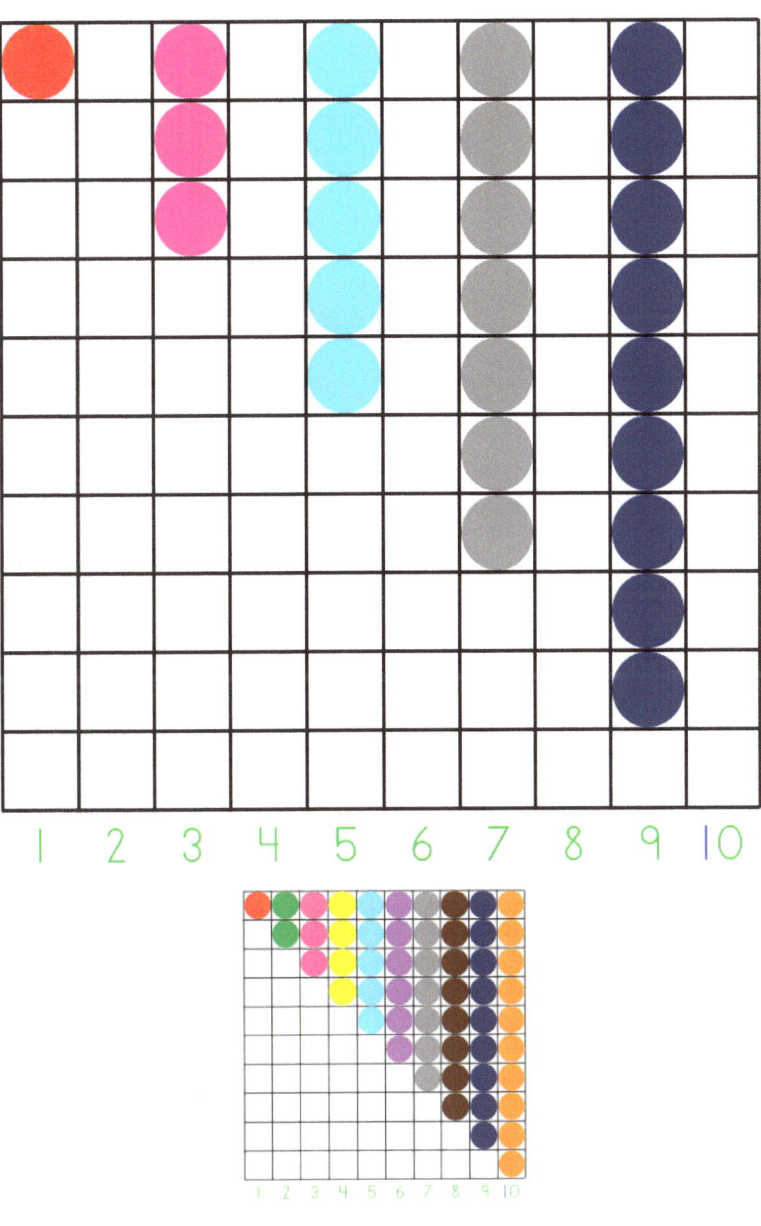

Exercise 1-2

Fill in the open spaces with the missing colored bead bars (or <u>quantities</u>) in their vertical arrangement.

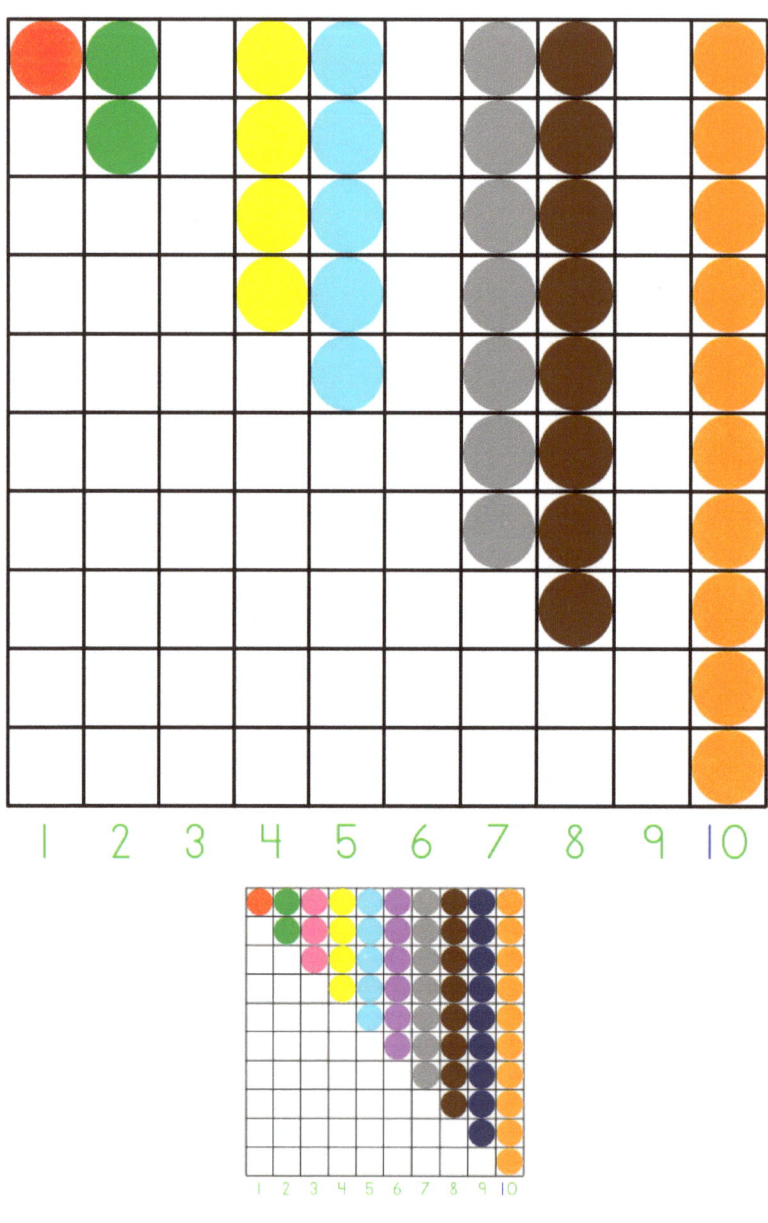

A Vision for Intellectual Wellness
In tribute to Dionicious and Mildred Gunawardena

Exercise 1-3

Fill in the open spaces with the missing colored bead bars (or <u>quantities</u>) in their vertical arrangement.

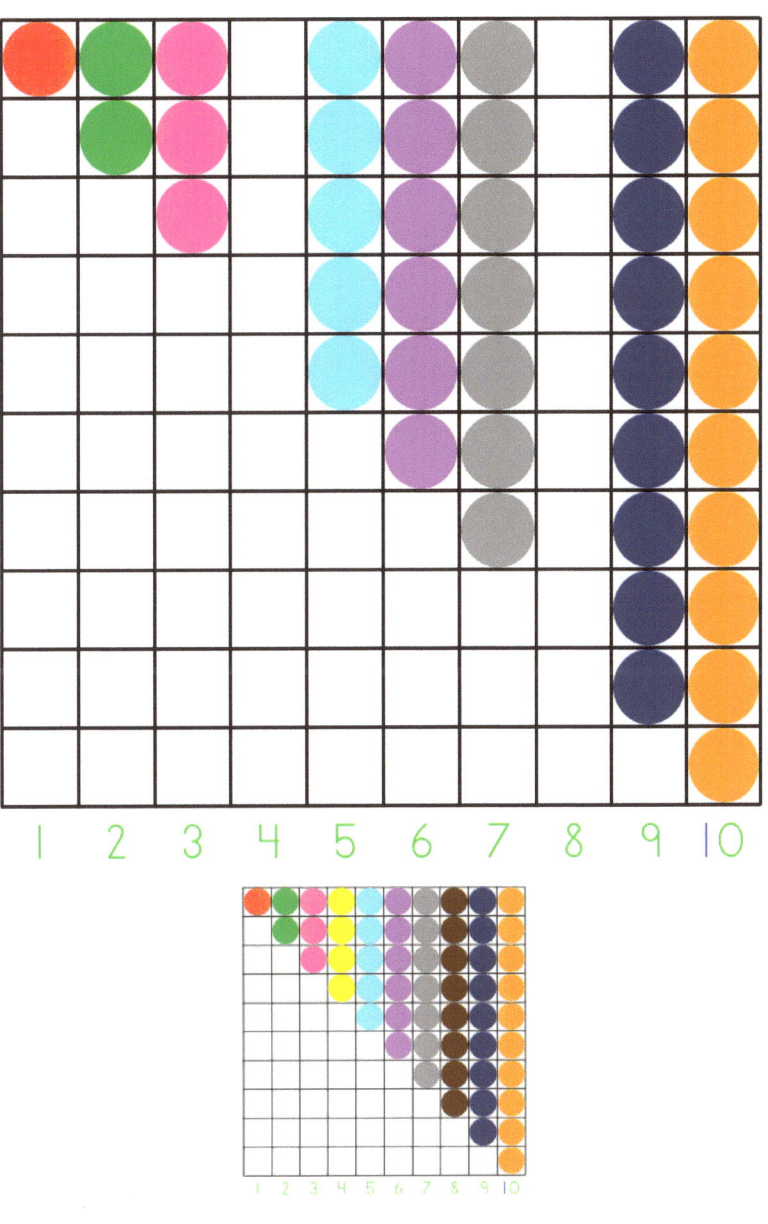

Exercise 1-4

Fill in the open spaces with the missing colored bead bars (or <u>quantities</u>) in their vertical arrangement.

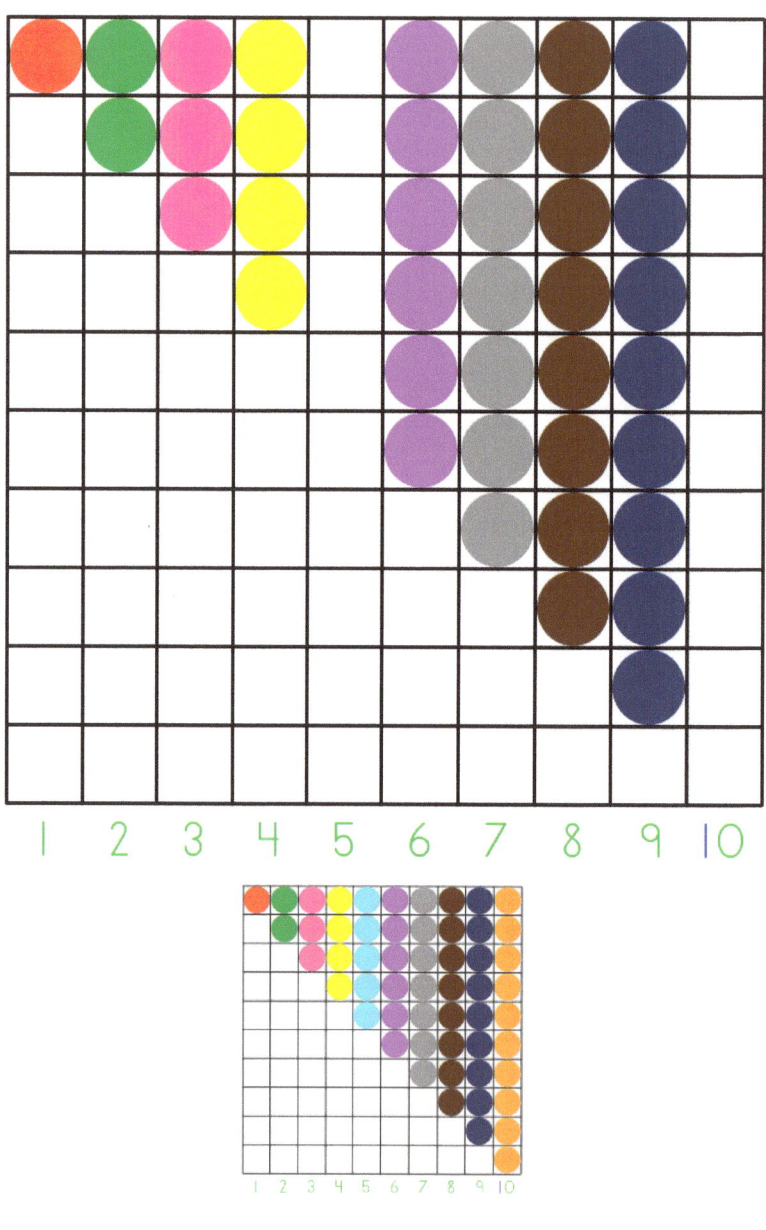

A Vision for Intellectual Wellness
In tribute to Dionicious and Mildred Gunawardena

Exercise 1-5

Fill in the open spaces with the missing colored bead bars (or <u>quantities</u>) 1 – 9 in their vertical arrangement, in sequence from left to right.

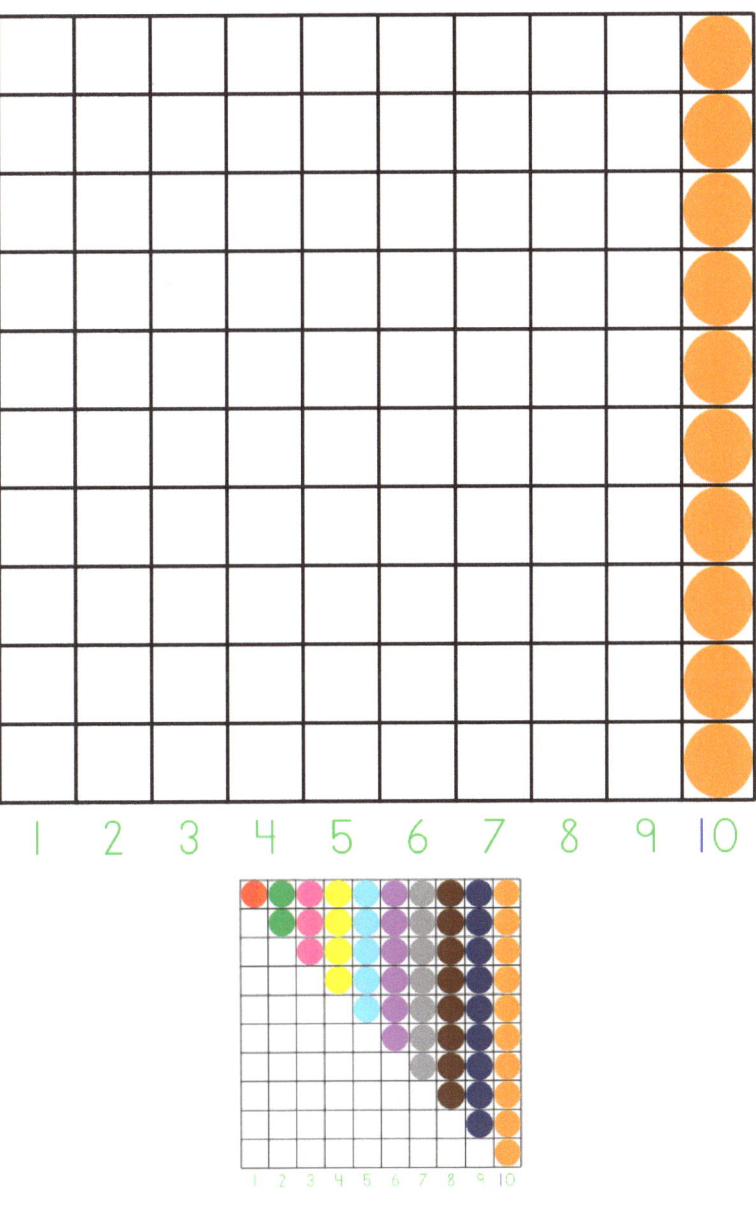

Exercise 1-6

Fill in the open spaces with the missing colored bead bars (or <u>quantities</u>) in their vertical arrangement <u>and</u> write their <u>numerical symbols</u>.

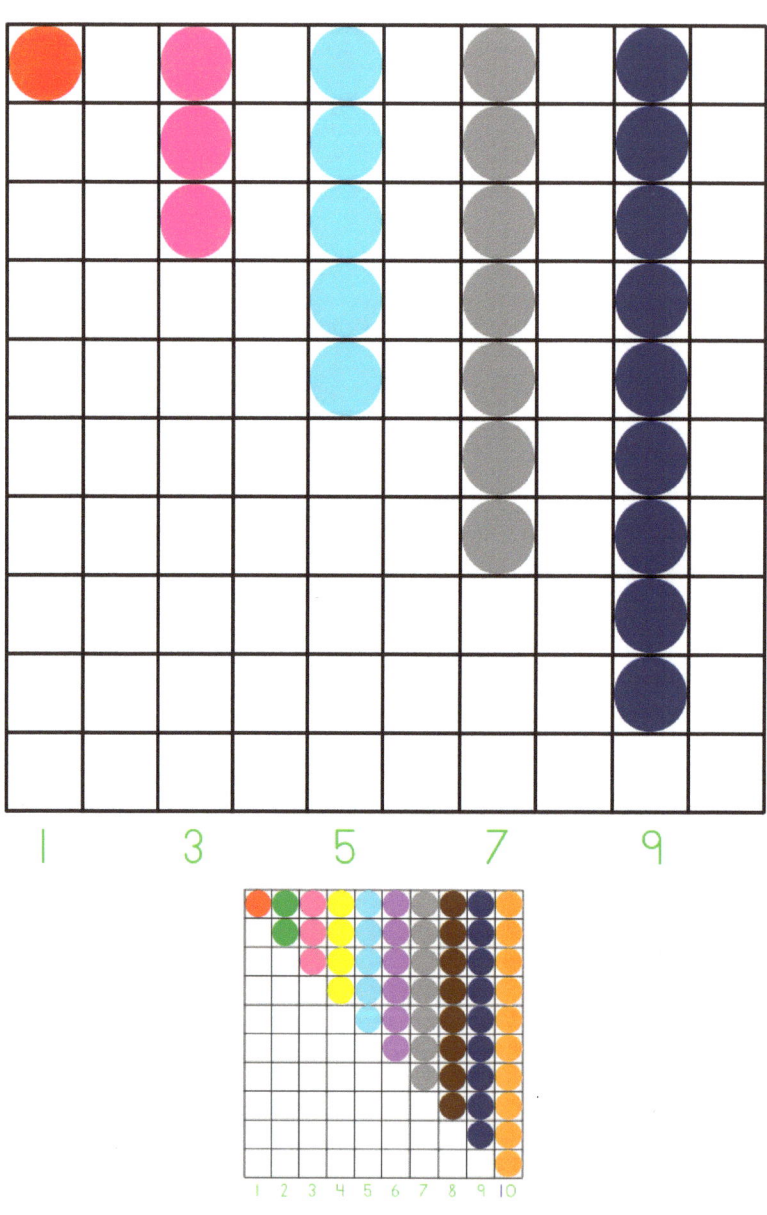

A Vision for Intellectual Wellness
In tribute to Dionicious and Mildred Gunawardena

Exercise 1-7

Fill in the open spaces with the missing colored bead bars (or <u>quantities</u>) in their vertical arrangement <u>and</u> write their <u>numerical symbols</u>.

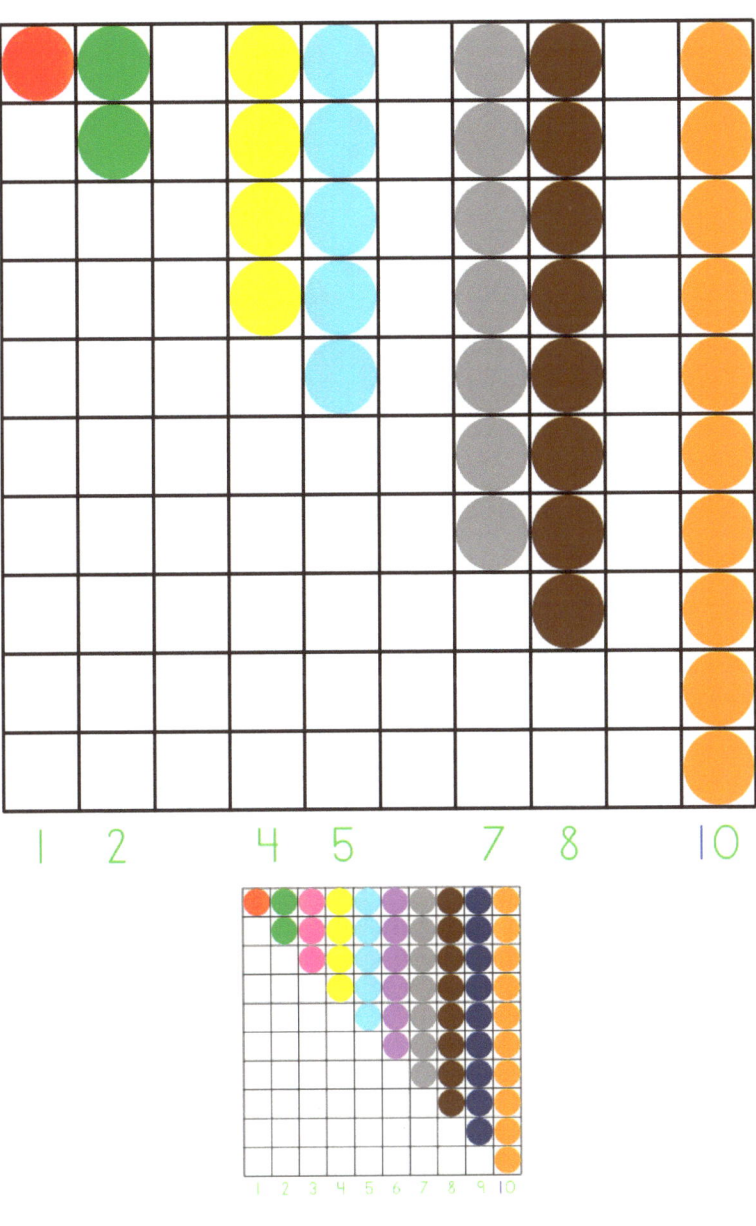

A Vision for Intellectual Wellness
In tribute to Dionicious and Mildred Gunawardena

Exercise 1-8

Fill in the open spaces with the missing colored bead bars (or <u>quantities</u>) in their vertical arrangement <u>and</u> write their <u>numerical symbols</u>.

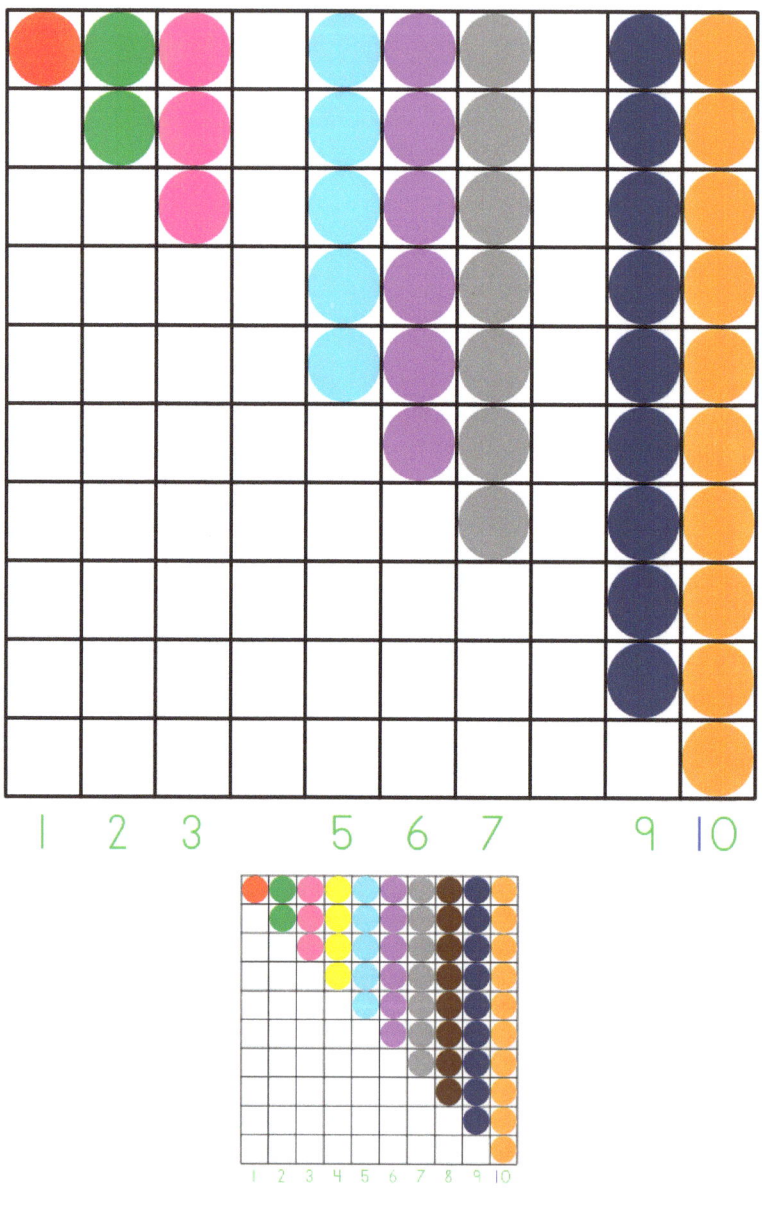

1 2 3   5 6 7   9 10

A Vision for Intellectual Wellness
In tribute to Dionicious and Mildred Gunawardena

Exercise 1-9

Fill in the open spaces with the missing colored bead bars (or <u>quantities</u>) in their vertical arrangement <u>and</u> write their <u>numerical symbols</u>.

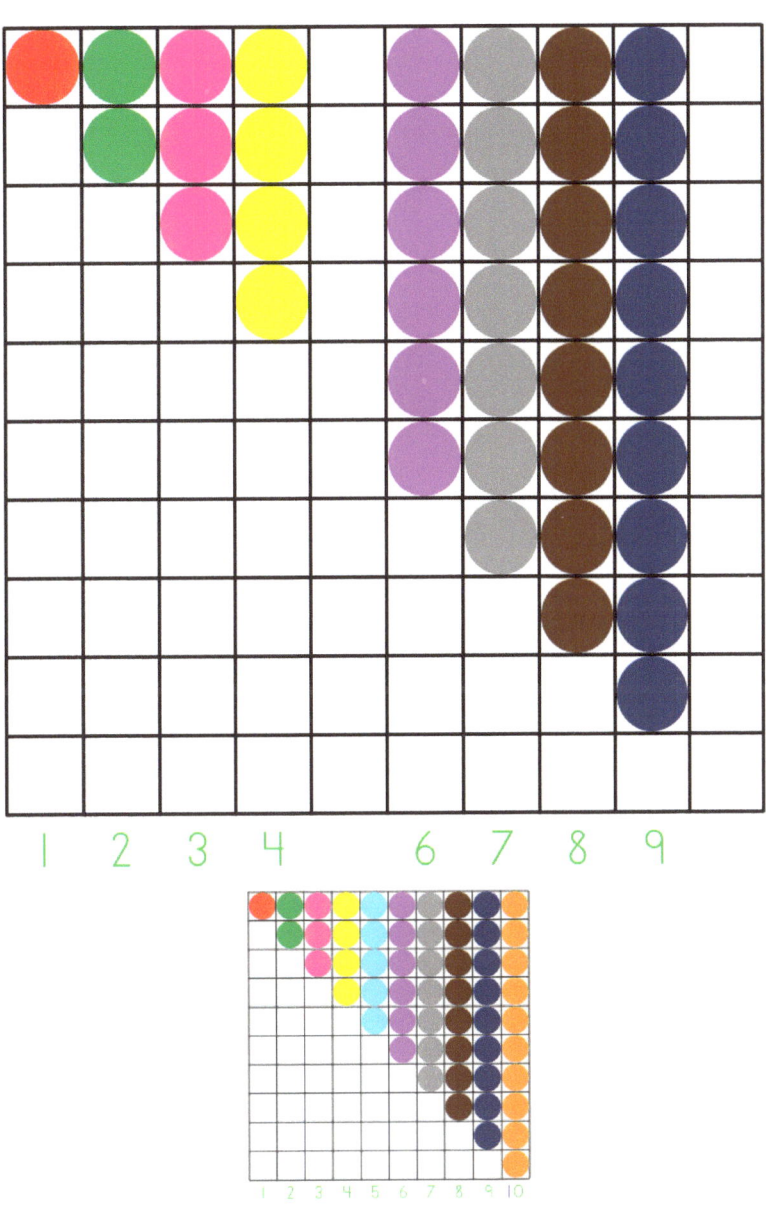

Exercise 1-10

Fill in the open spaces with the missing colored bead bars (or <u>quantities</u>) in their vertical arrangement <u>and</u> write their <u>numerical symbols</u> from 1 – 9 in sequence, from left to right.

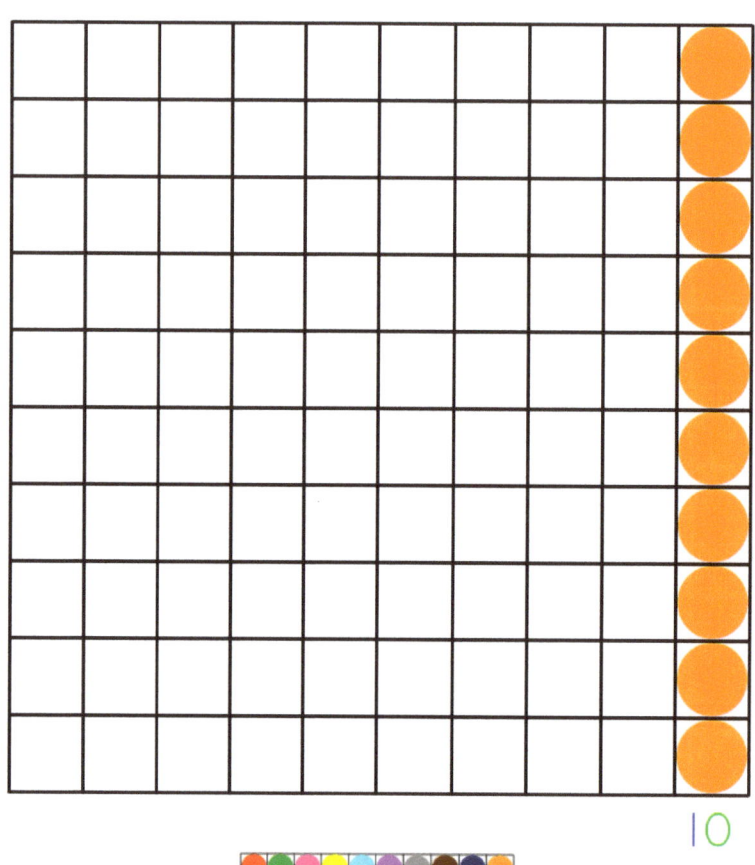

10

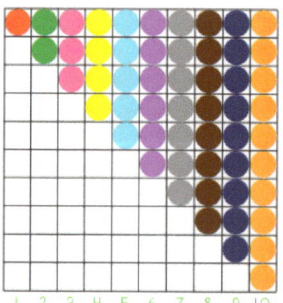

Section 2

<u>Cardinal</u> Numbers
in Horizontal Order

Presentation Chart 2

<u>Cardinal</u> Numbers in Horizontal Order

Illustrated below are the <u>quantities</u> and associated <u>numerical symbols</u> of <u>cardinal numbers</u>. (Note the vocabulary underlined.)

- Observe the quantities arranged along a horizontal sequence.
- Observe the numerals associated with the quantities.
- Observe the particular color assigned to each quantity.

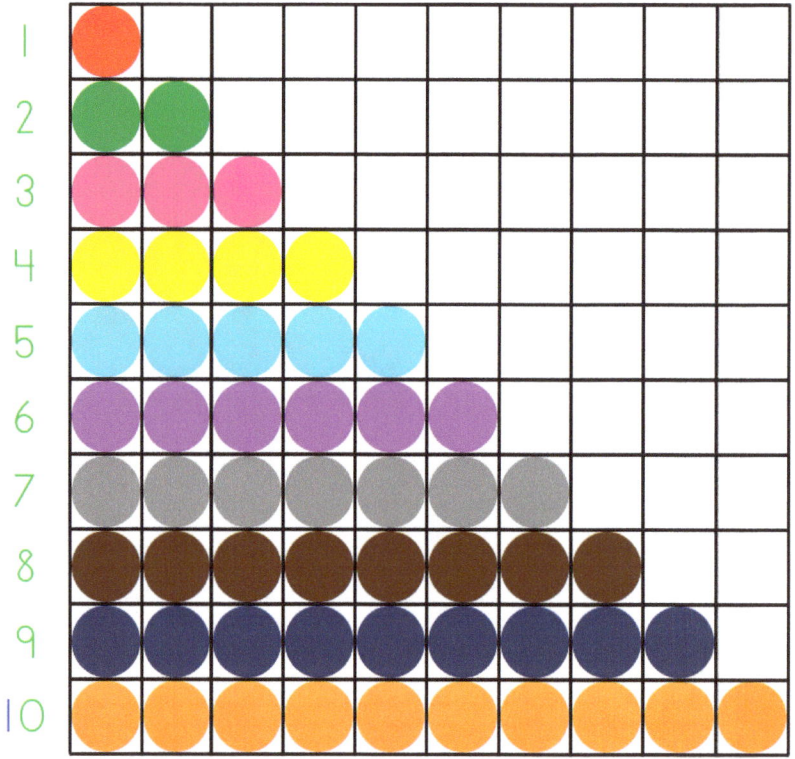

Exercise 2-1

Fill in the open spaces with the missing colored bead bars (or <u>quantities</u>) in their horizontal arrangement.

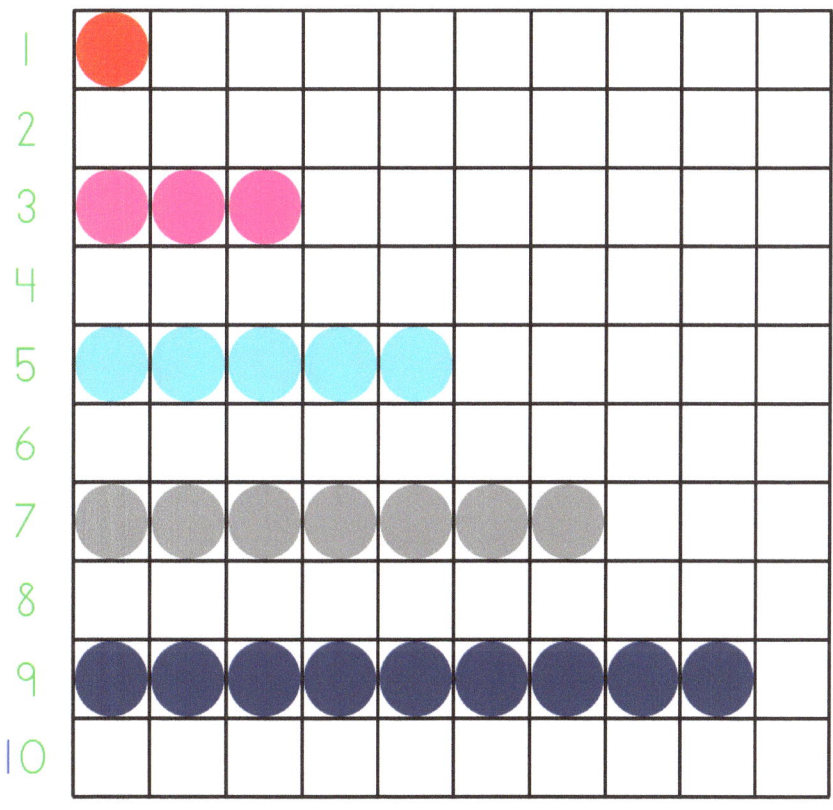

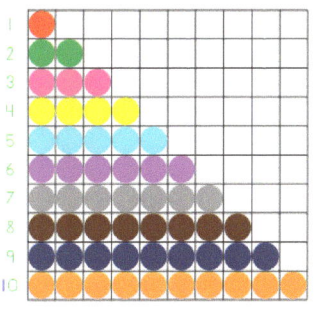

Exercise 2-2

Fill in the open spaces with the missing colored bead bars (or <u>quantities</u>) in their horizontal arrangement.

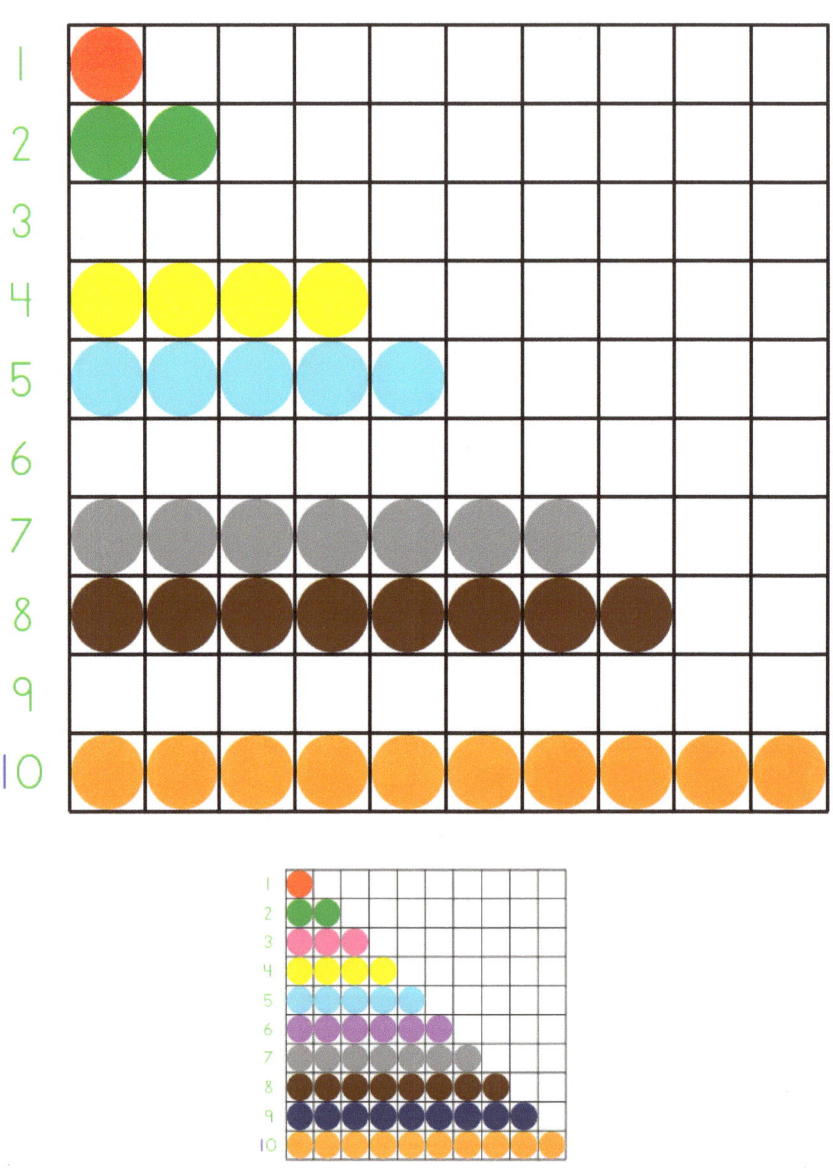

A Vision for Intellectual Wellness
In tribute to Dionicious and Mildred Gunawardena

Exercise 2-3

Fill in the open spaces with the missing colored bead bars (or <u>quantities</u>) in their horizontal arrangement.

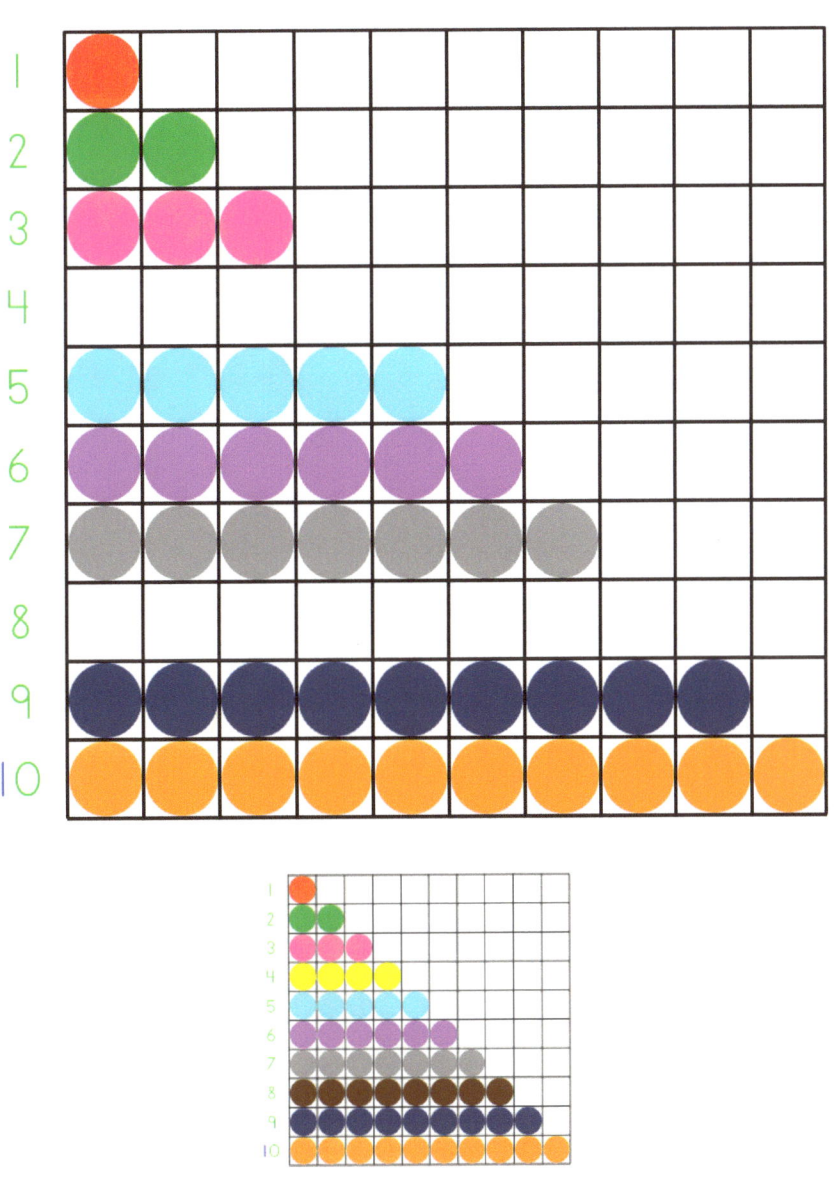

Exercise 2-4

Fill in the open spaces with the missing colored bead bars (or <u>quantities</u>) in their horizontal arrangement.

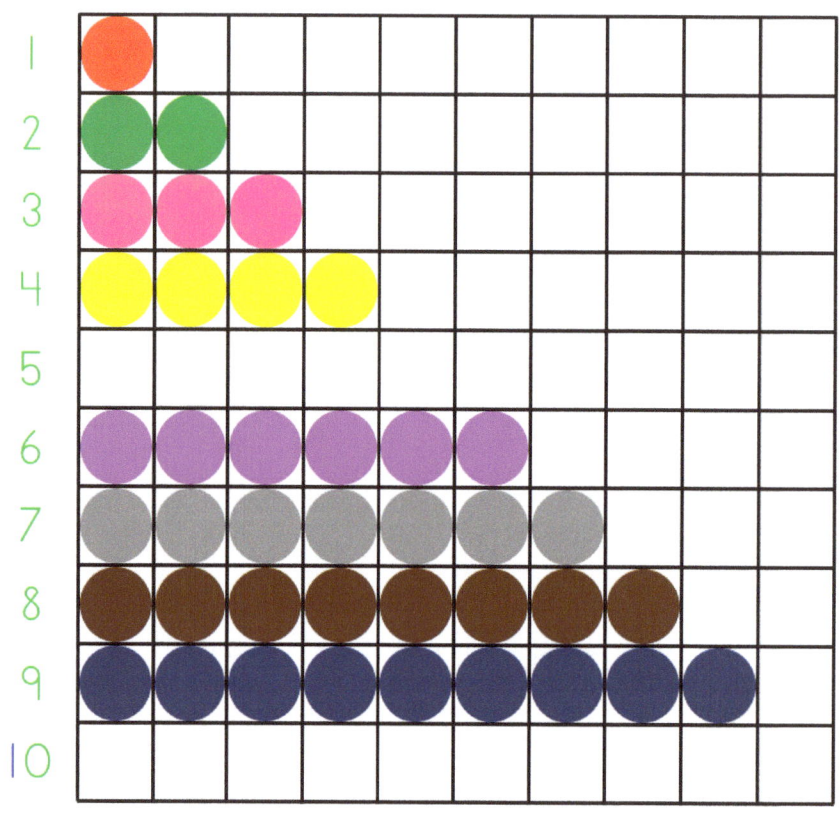

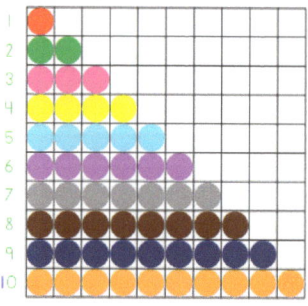

A Vision for Intellectual Wellness
In tribute to Dionicious and Mildred Gunawardena

Exercise 2-5

Fill in the open spaces with the missing colored bead bars (or <u>quantities</u>)   1 – 9 in their horizontal arrangement, in sequence from top to bottom.

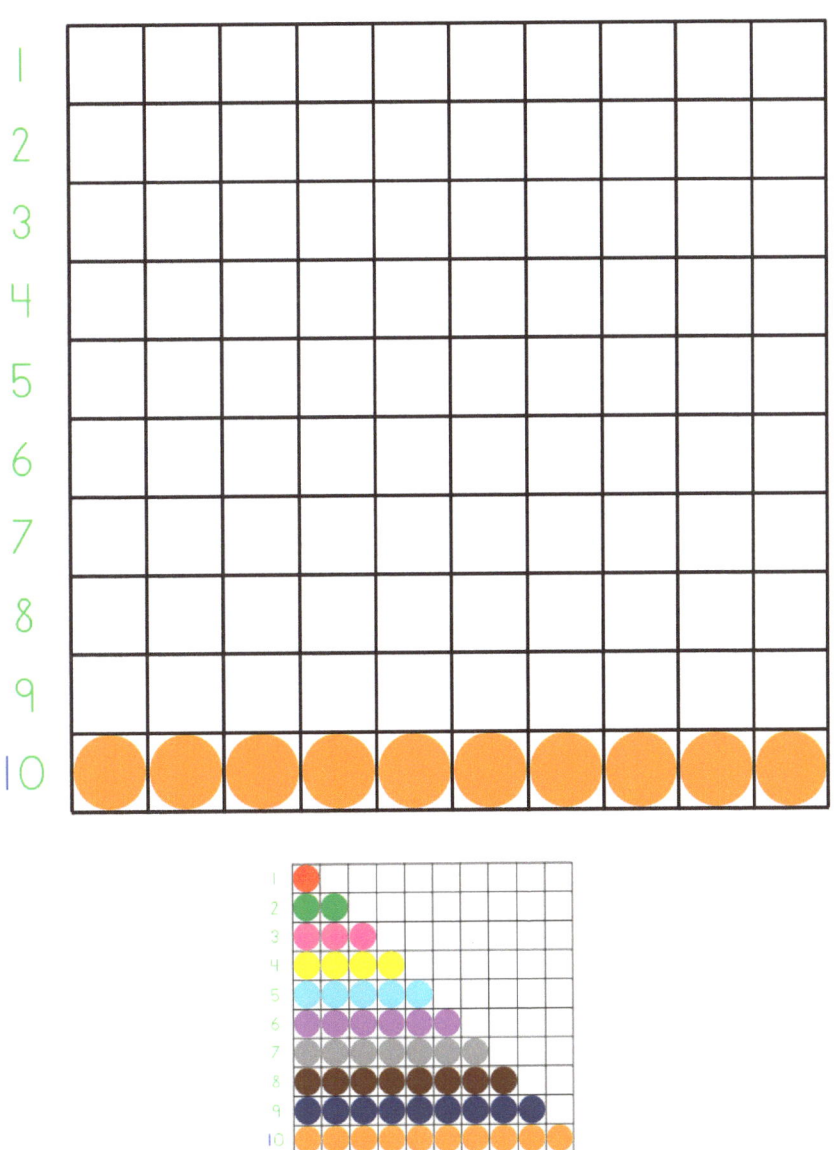

Exercise 2-6

Fill in the open spaces with the missing colored bead bars (or <u>quantities</u>) in their horizontal arrangement <u>and</u> write their <u>numerical symbols</u>.

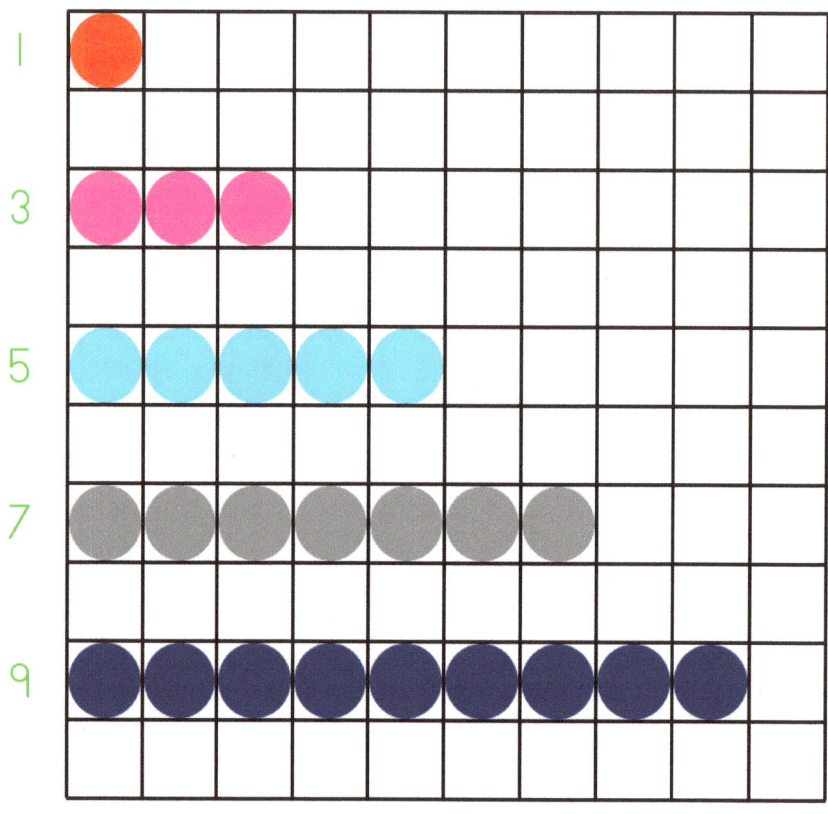

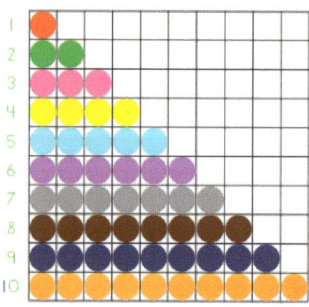

Exercise 2-7

Fill in the open spaces with the missing colored bead bars (or <u>quantities</u>) in their horizontal arrangement <u>and</u> write their <u>numerical symbols</u>.

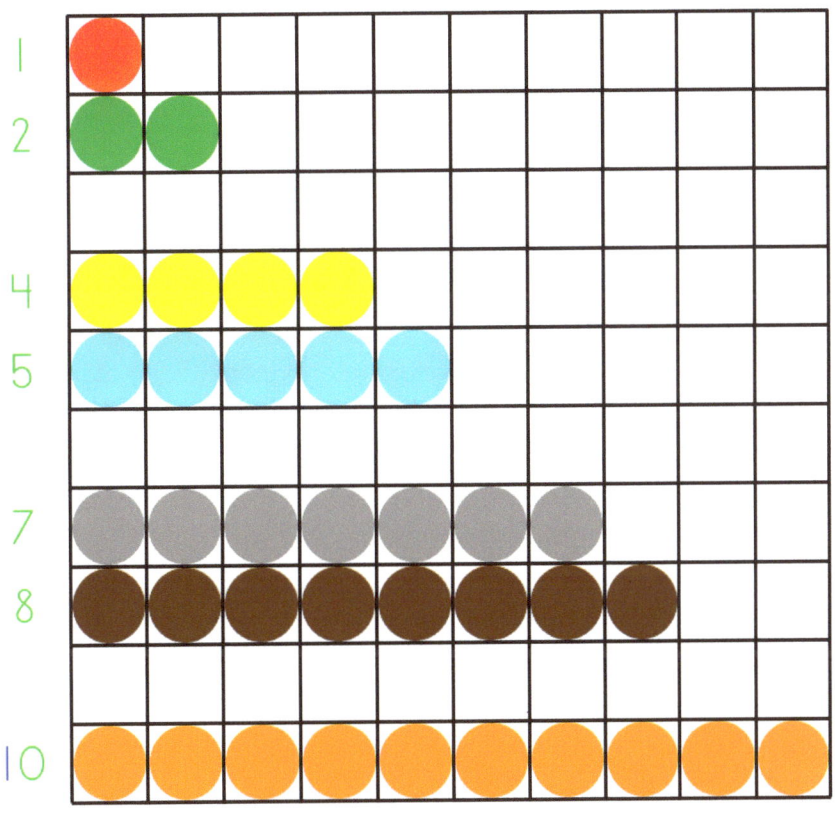

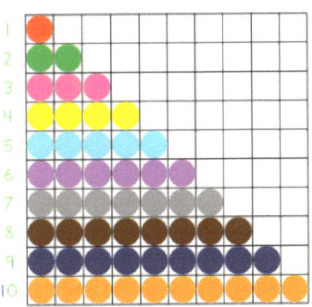

Exercise 2-8

Fill in the open spaces with the missing colored bead bars (or <u>quantities</u>) in their horizontal arrangement <u>and</u> write their <u>numerical symbols</u>.

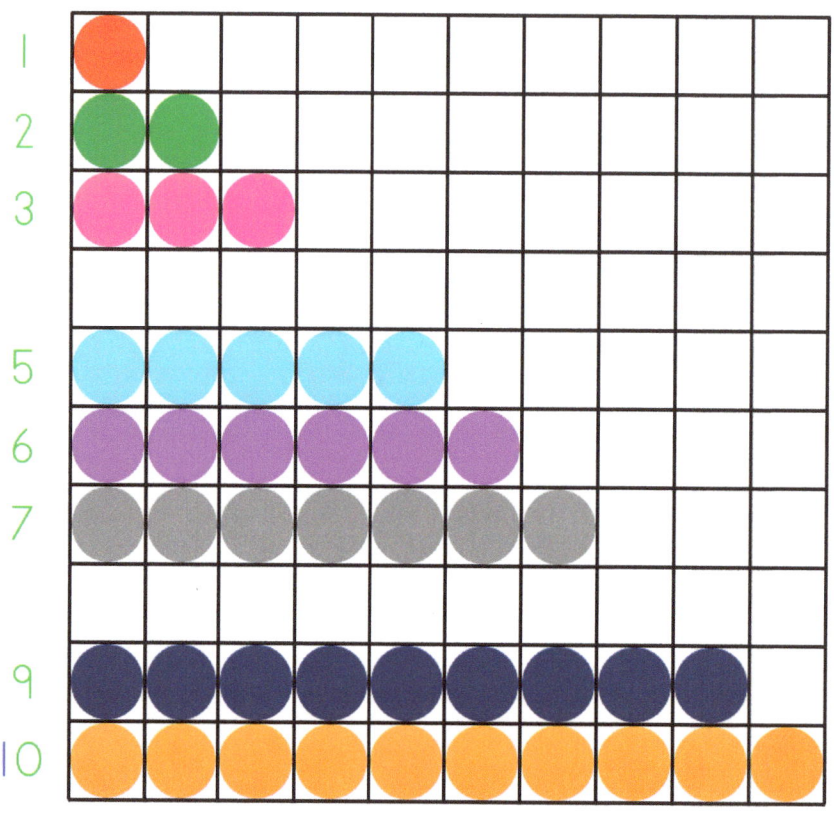

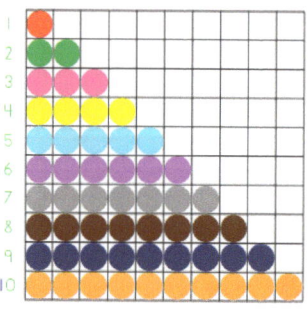

A Vision for Intellectual Wellness
In tribute to Dionicious and Mildred Gunawardena

Exercise 2-9

Fill in the open spaces with the missing colored bead bars (or <u>quantities</u>) in their horizontal arrangement <u>and</u> write their <u>numerical symbols</u>.

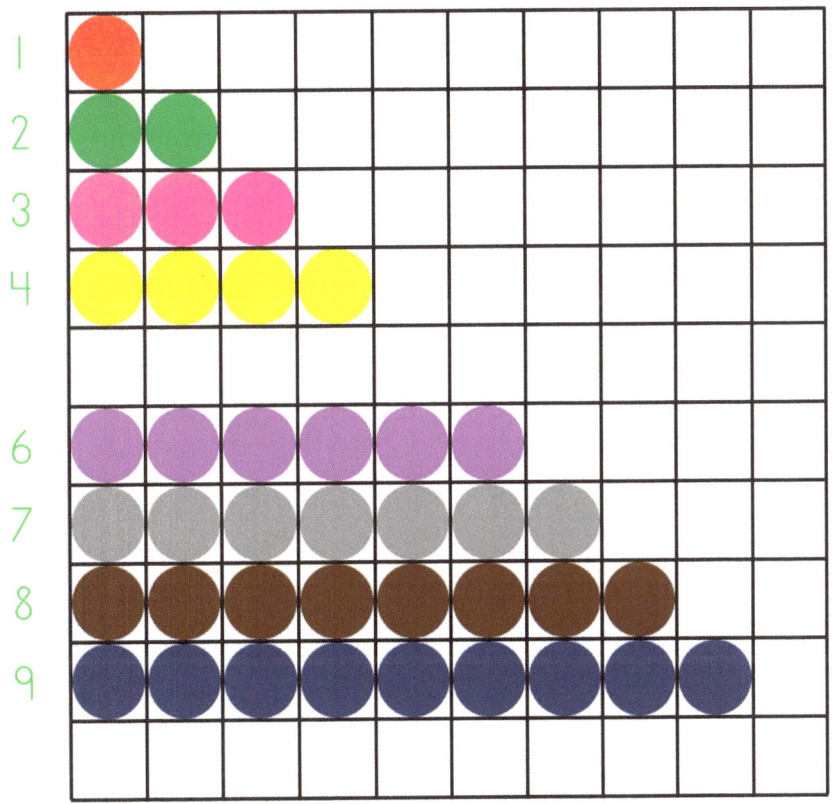

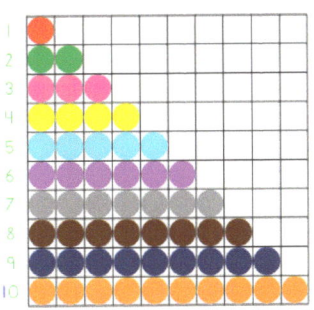

Exercise 2-10

Fill in the open spaces with the missing colored bead bars (or <u>quantities</u>) in their horizontal arrangement <u>and</u> write their <u>numerical symbols</u> from 1 – 9 in sequence, from top to bottom.

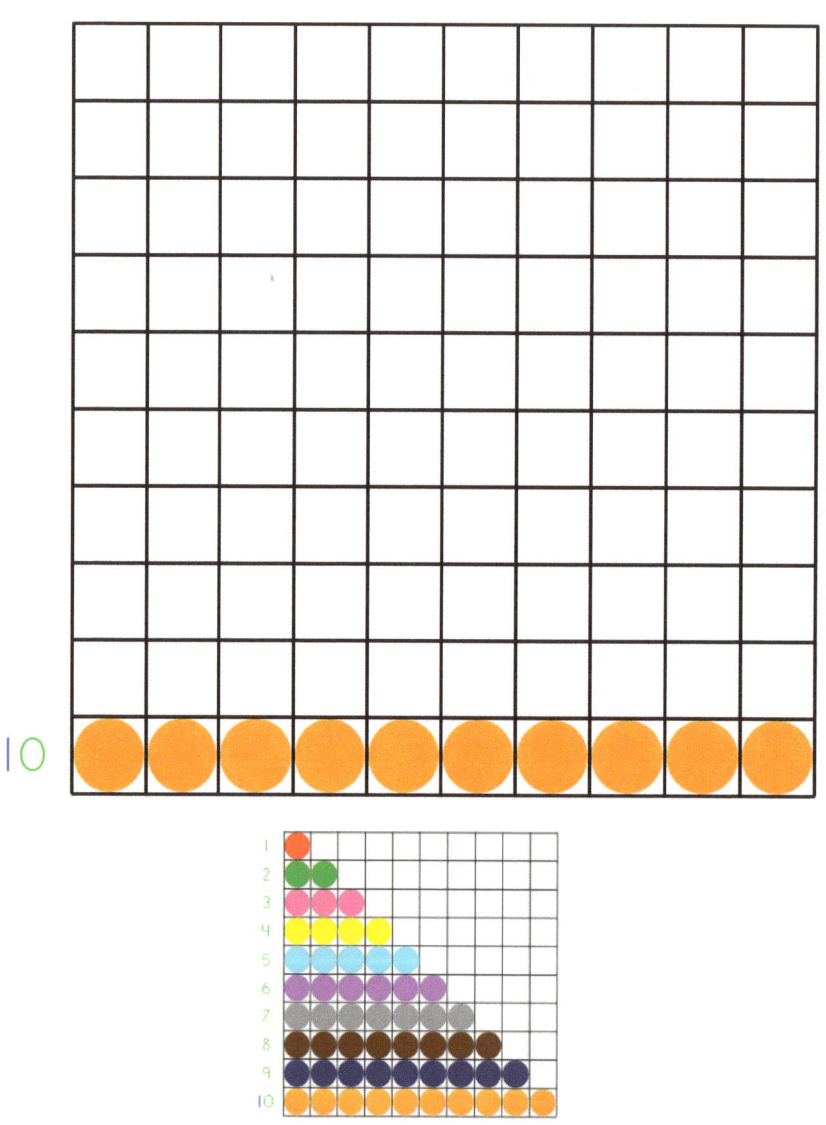

Review Exercise A – <u>Cardinal</u> Numbers
1-4 Vertical

Fill in the open spaces with the missing colored bead bars (or <u>quantities</u>) in their vertical arrangement <u>and</u> write their <u>numerical symbols</u>.

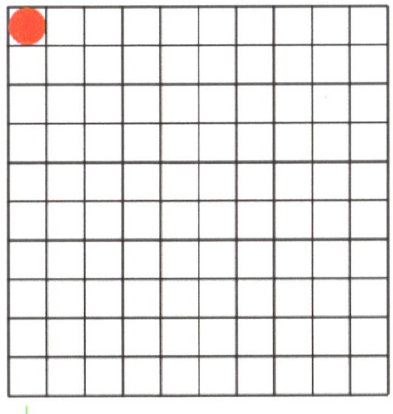

1

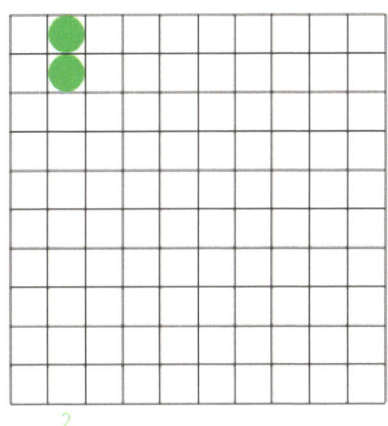

2

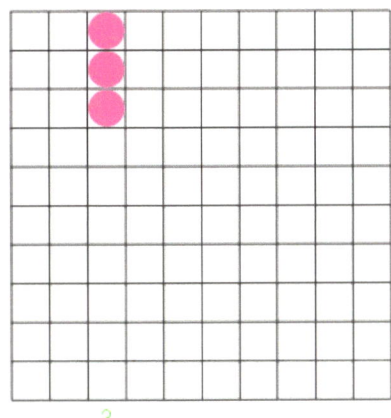

3

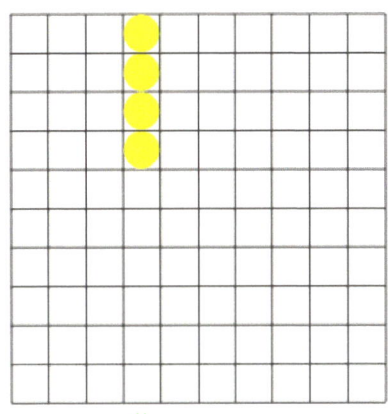

4

Review Exercise A – <u>Cardinal</u> Numbers
5-8 Vertical

Fill in the open spaces with the missing colored bead bars (or <u>quantities</u>) in their vertical arrangement <u>and</u> write their <u>numerical symbols</u>.

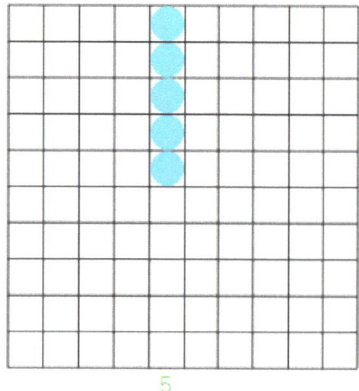

5

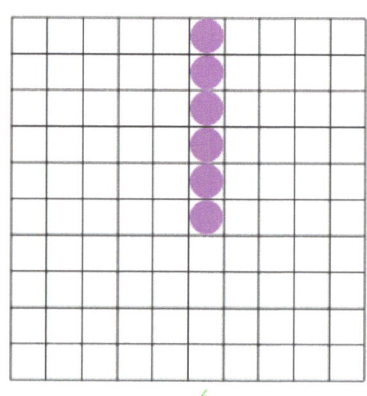

6

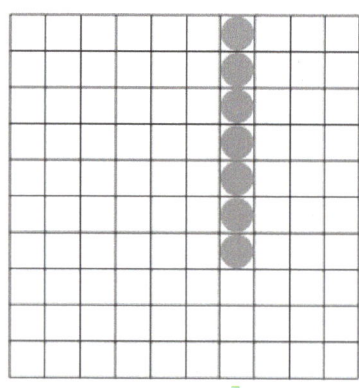

7

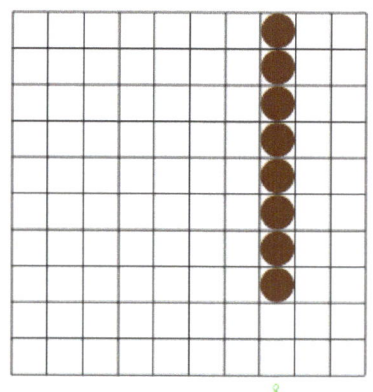

8

A Vision for Intellectual Wellness
In tribute to Dionicious and Mildred Gunawardena

Review Exercise A – <u>Cardinal</u> Numbers
9-10 Vertical

Fill in the open spaces with the missing colored bead bars (or <u>quantities</u>) in their vertical arrangement <u>and</u> write their <u>numerical symbols</u>.

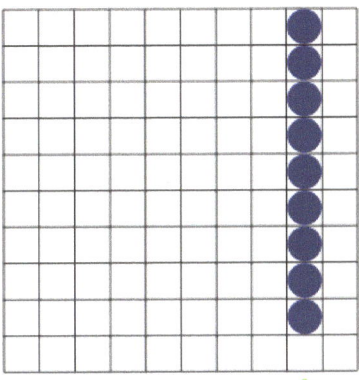

9

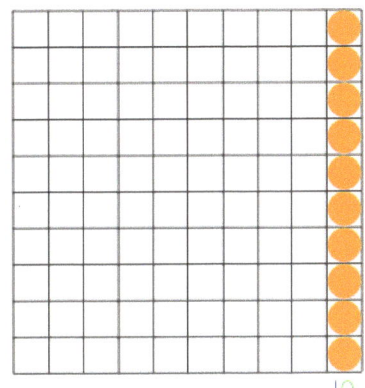

10

Review Exercise B – <u>Cardinal</u> Numbers
1-4 Horizontal

Fill in the open spaces with the missing colored bead bars (or <u>quantities</u>) in their horizontal arrangement <u>and</u> write their <u>numerical symbols</u>.

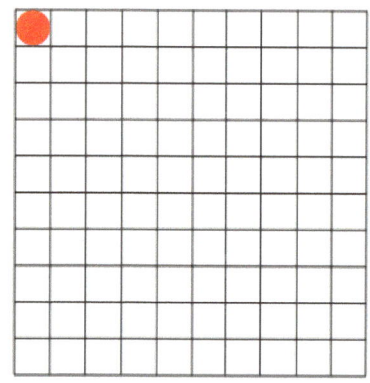

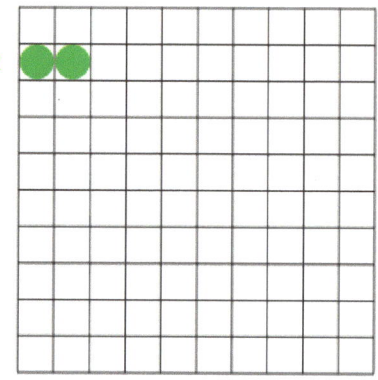

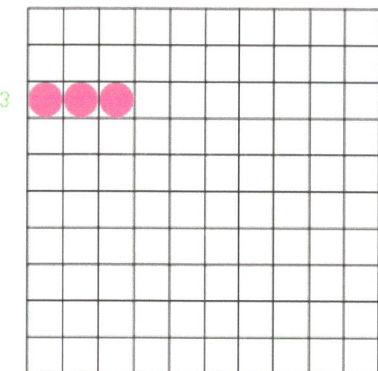

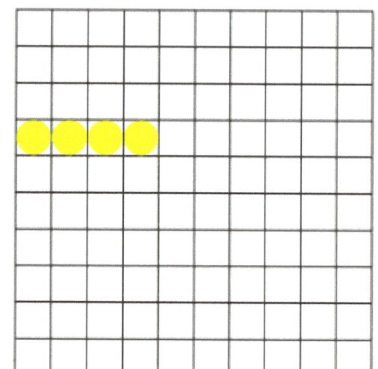

A Vision for Intellectual Wellness
In tribute to Dionicious and Mildred Gunawardena

Review Exercise B – <u>Cardinal</u> Numbers
5-8 Horizontal

Fill in the open spaces with the missing colored bead bars (or <u>quantities</u>) in their horizontal arrangement <u>and</u> write their <u>numerical symbols</u>.

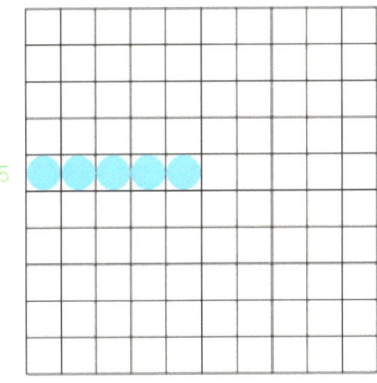

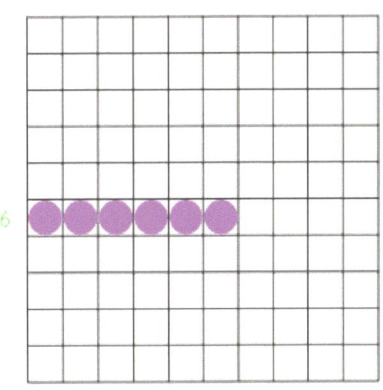

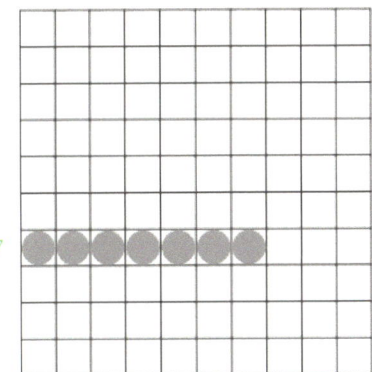

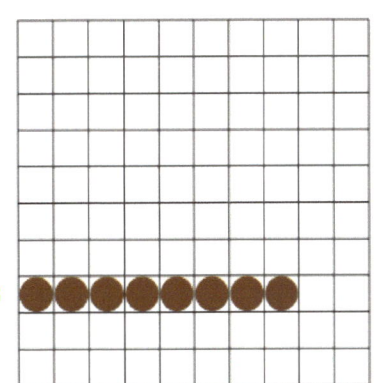

Review Exercise B – <u>Cardinal</u> Numbers
9-10 Horizontal

Fill in the open spaces with the missing colored bead bars (or <u>quantities</u>) in their horizontal arrangement <u>and</u> write their <u>numerical symbols</u>.

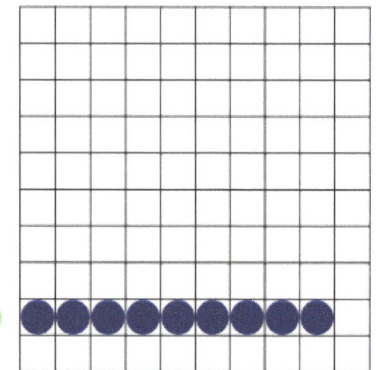 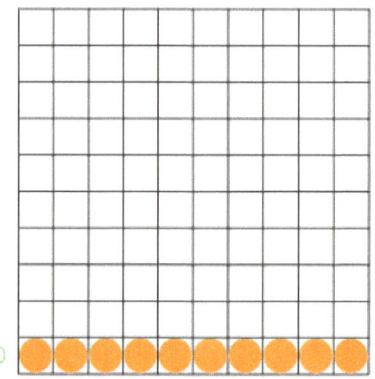

A Vision for Intellectual Wellness
In tribute to Dionicious and Mildred Gunawardena

Review Exercise C – <u>Cardinal</u> Numbers
1-10 Vertical and Horizontal Quantities Converge to Form Squares

Observe the pattern of colored bead bars (or <u>quantities</u>) arranged in their simultaneously vertical and horizontal sequence.

EXAMPLE:

An original from The Precious Jewels of Mrs. "G"
Basic Number Concepts 1-9

Review Exercise C – <u>Cardinal</u> Numbers
1-4 Vertical and Horizontal Quantities Converge to Form Squares

Fill in the open spaces with the missing colored bead bars (or <u>quantities</u>) in their vertical and horizontal arrangement <u>and</u> write their <u>numerical symbols</u>.

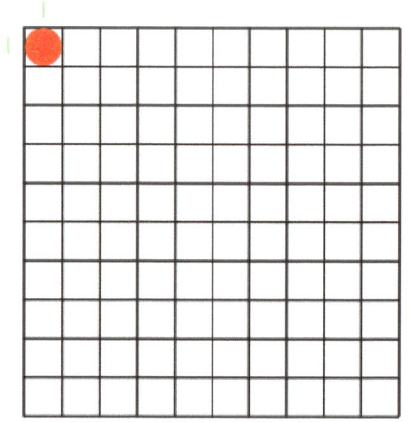
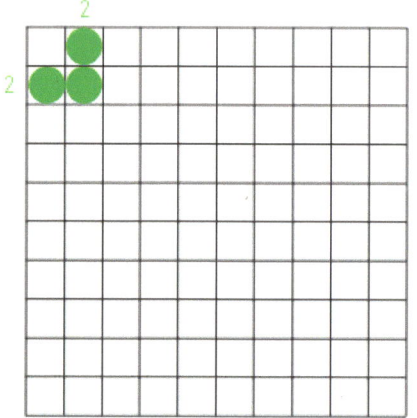
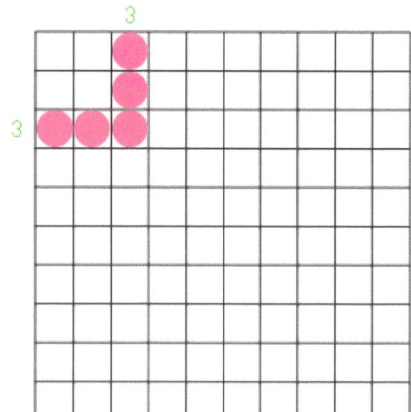
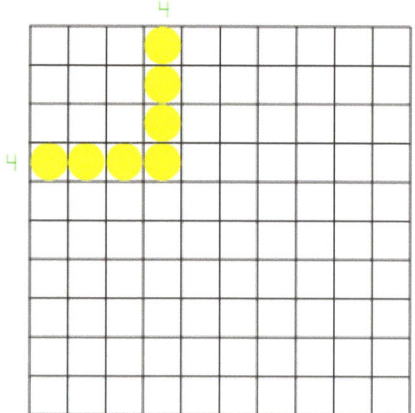

A Vision for Intellectual Wellness
In tribute to Dionicious and Mildred Gunawardena

Review Exercise C – <u>Cardinal</u> Numbers
5-8 Vertical and Horizontal Quantities Converge to Form Squares

Fill in the open spaces with the missing colored bead bars (or <u>quantities</u>) in their vertical and horizontal arrangement <u>and</u> write their <u>numerical symbols</u>.

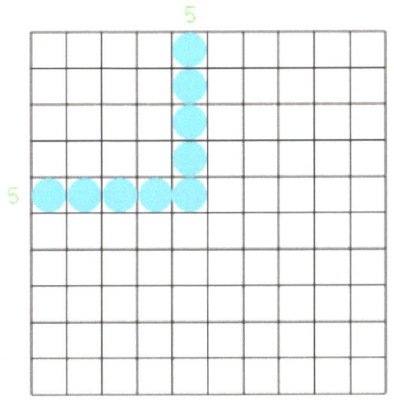

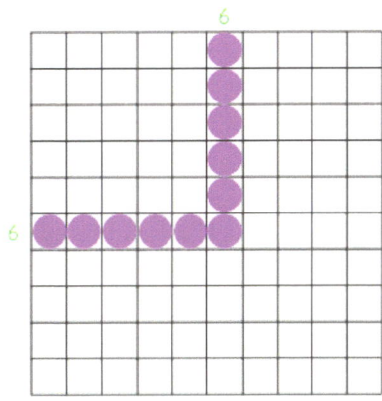

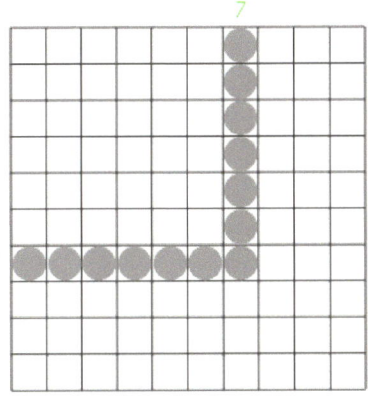

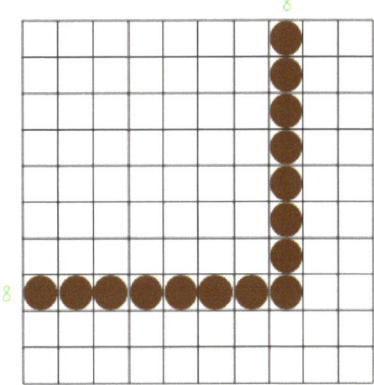

Review Exercise C – <u>Cardinal</u> Numbers
9-10 Vertical and Horizontal Quantities Converge to Form Squares

Fill in the open spaces with the missing colored bead bars (or <u>quantities</u>) in their vertical and horizontal arrangement <u>and</u> write their <u>numerical symbols</u>.

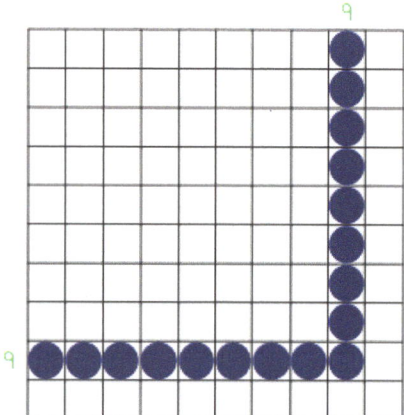

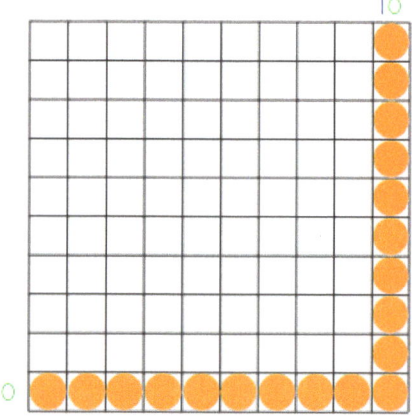

A Vision for Intellectual Wellness
In tribute to Dionicious and Mildred Gunawardena

Review Exercise D – <u>Cardinal</u> Numbers
1-10 Vertical and Horizontal Quantities Converge to Form Squares

Fill in the color defined quantities to match their associated numerals as shown in their vertical and horizontal sequence.

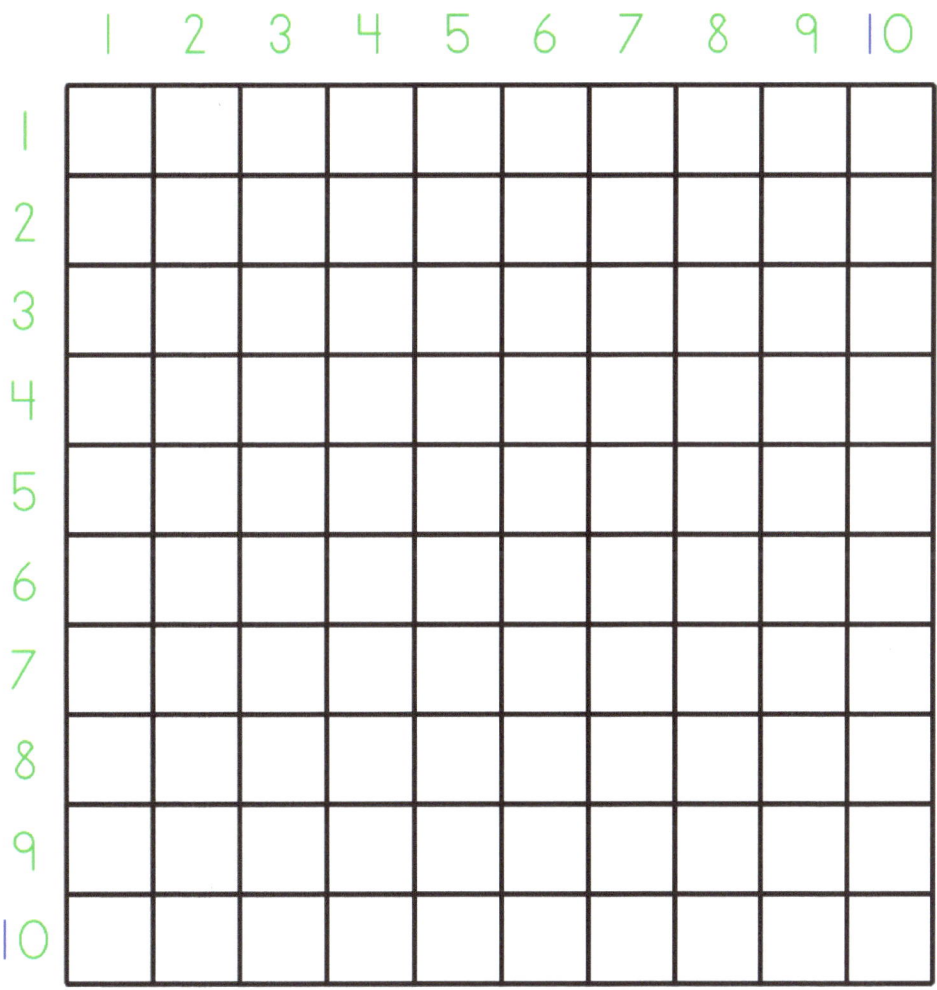

Review Exercise E – <u>Cardinal</u> Numbers
1-10 Vertical and Horizontal Quantities Converge to Form Squares

Fill in the color defined quantities in their vertical and horizontal sequence along with their associated numerals.

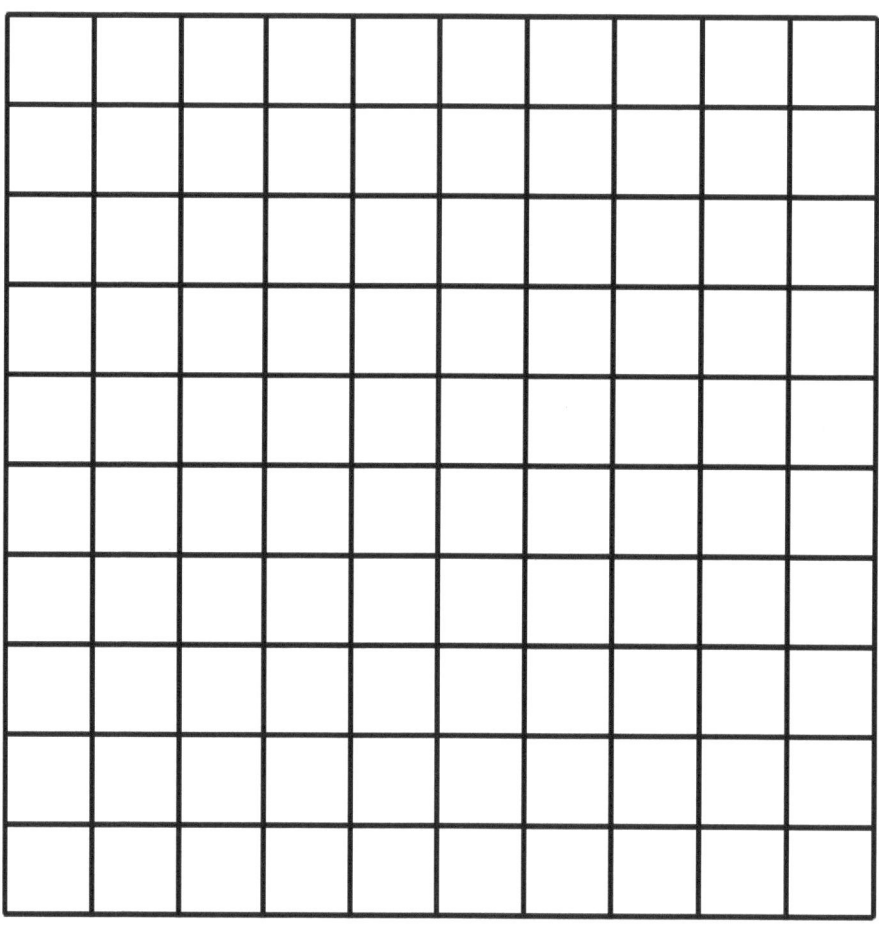

Section 3

<u>Ordinal</u> Numbers in Vertical Order

Presentation Chart 3

<u>Ordinal</u> Numbers in Vertical Order

Illustrated below are the <u>quantities</u> and their associated <u>sequential numeric positions</u> as <u>ordinal numbers</u>. (Note the vocabulary underlined.)

- Observe the quantities arranged along a vertical sequence.
- Observe the numerals associated with the quantities.
- Observe the particular color assigned to each quantity.

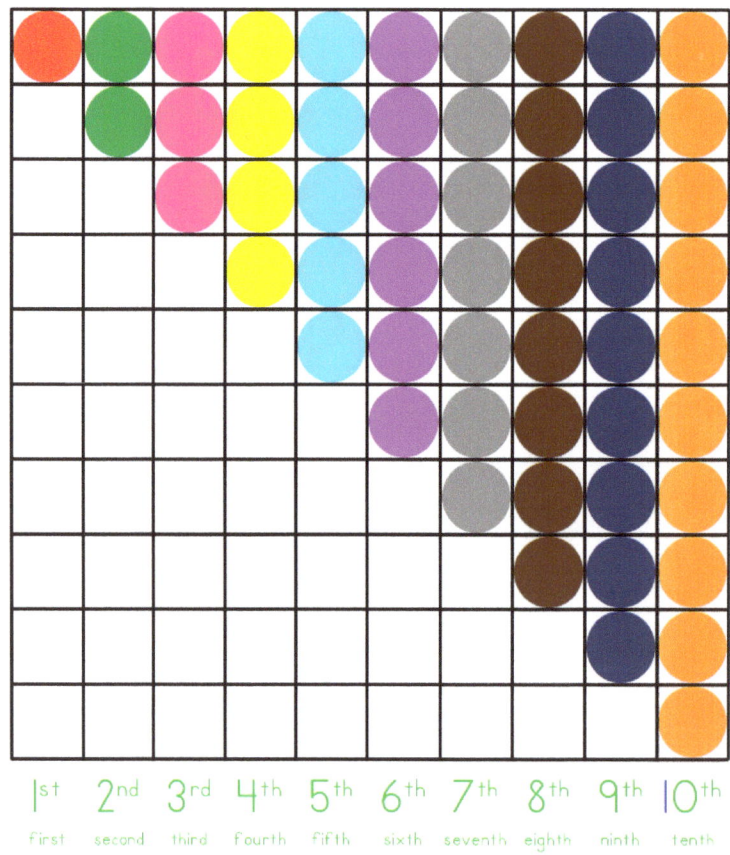

A Vision for Intellectual Wellness
In tribute to Dionicious and Mildred Gunawardena

Exercise 3-1

Fill in the open spaces with the missing colored bead bars (as <u>sequential numeric positions</u>) in their vertical arrangement.

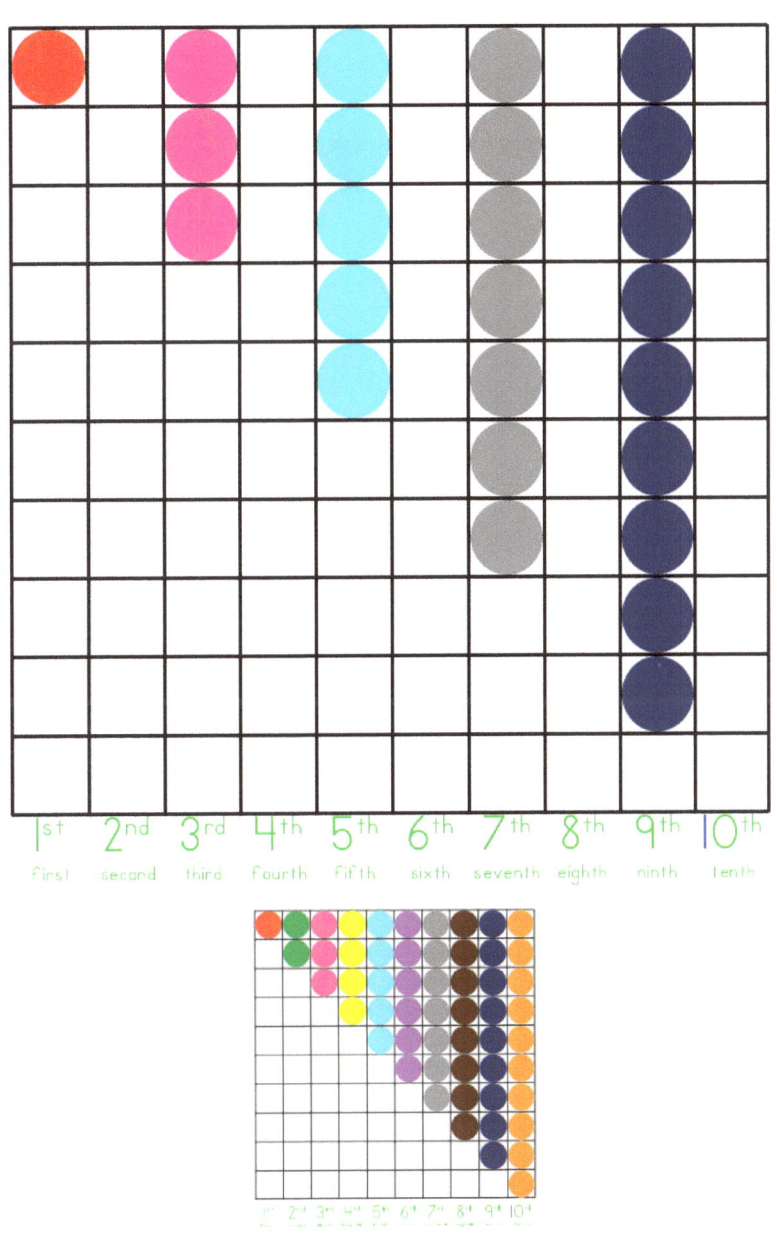

Exercise 3-2

Fill in the open spaces with the missing colored bead bars (as <u>sequential numeric positions</u>) in their vertical arrangement.

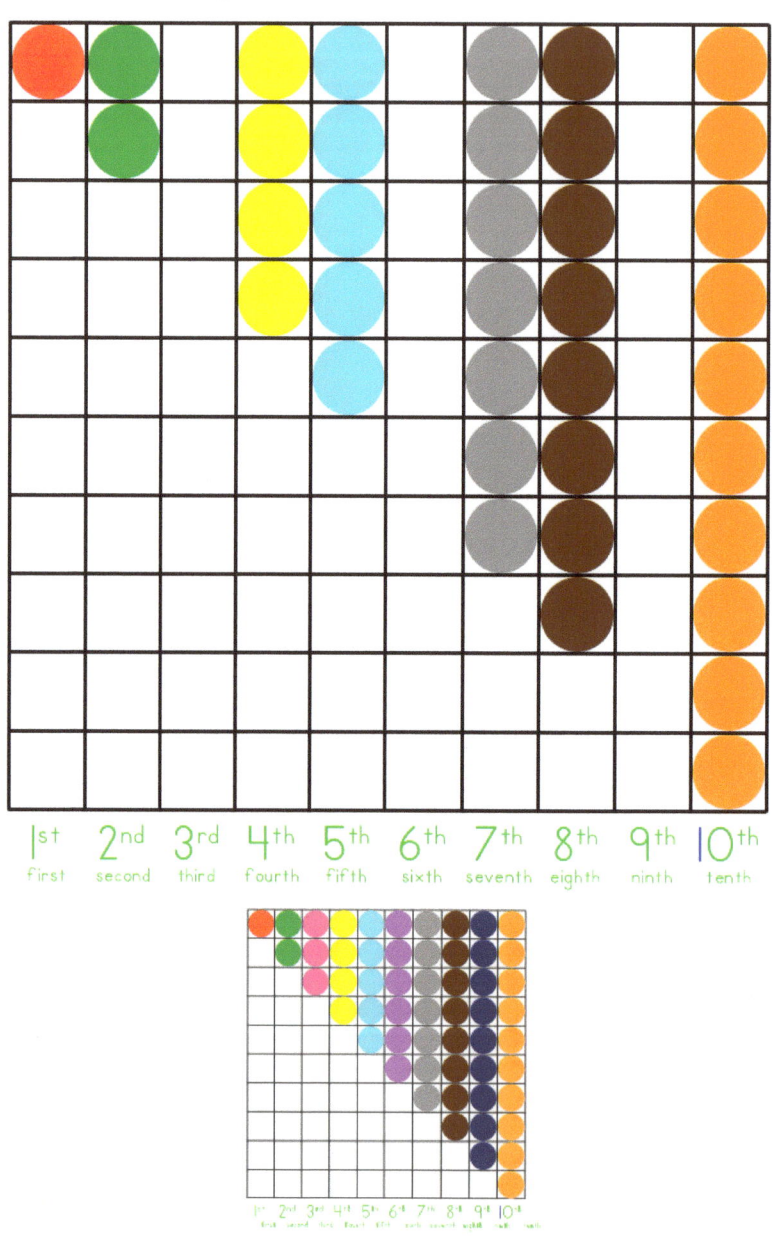

Exercise 3-3

Fill in the open spaces with the missing colored bead bars (as <u>sequential numeric positions</u>) in their vertical arrangement.

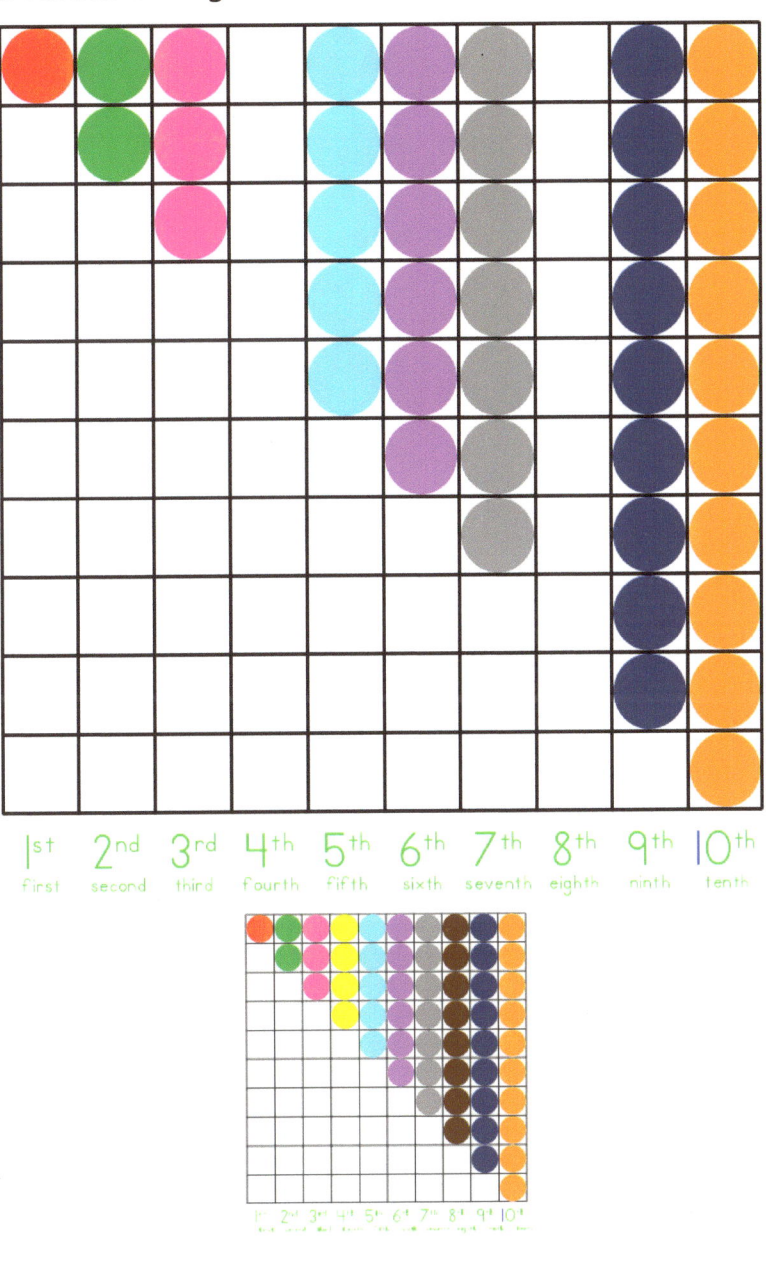

Exercise 3-4

Fill in the open spaces with the missing colored bead bars (as <u>sequential numeric positions</u>) in their vertical arrangement.

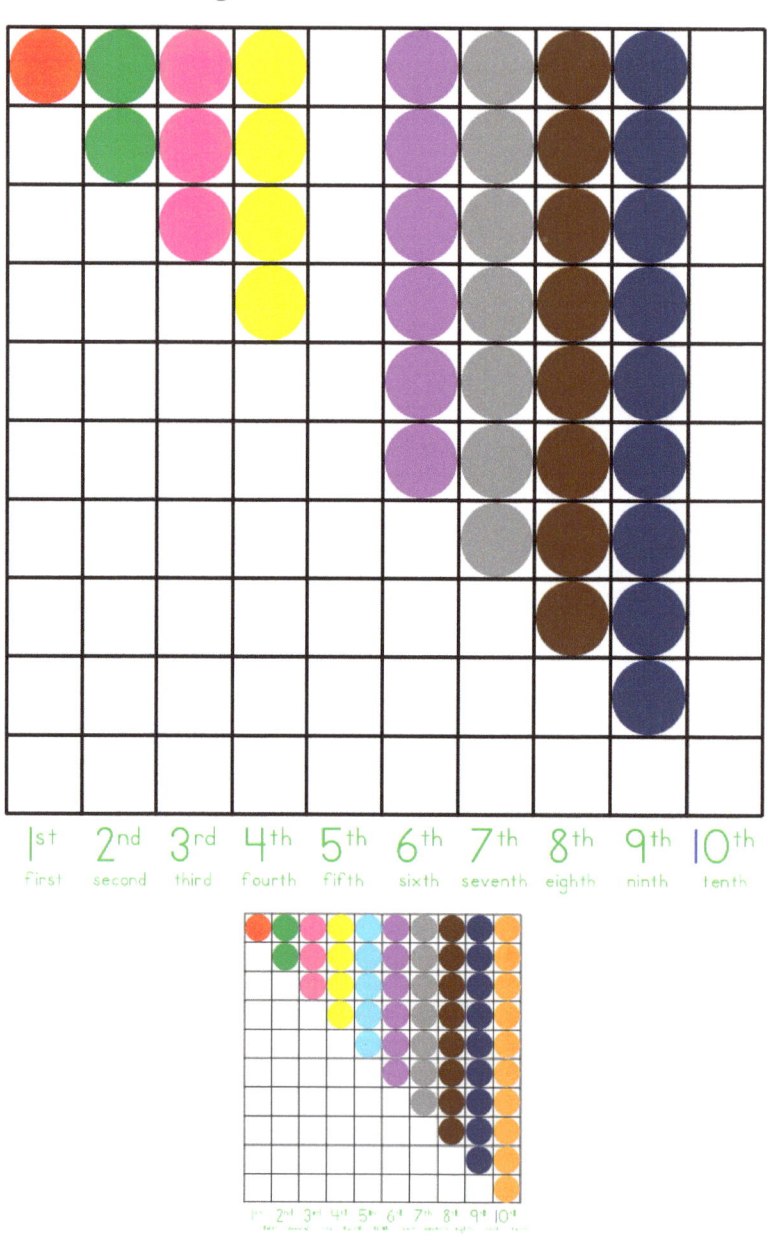

A Vision for Intellectual Wellness
In tribute to Dionicious and Mildred Gunawardena

Exercise 3-5

Fill in the open spaces with the missing colored bead bars (as <u>sequential numeric positions</u>) 1 – 9 in their vertical arrangement, from left to right.

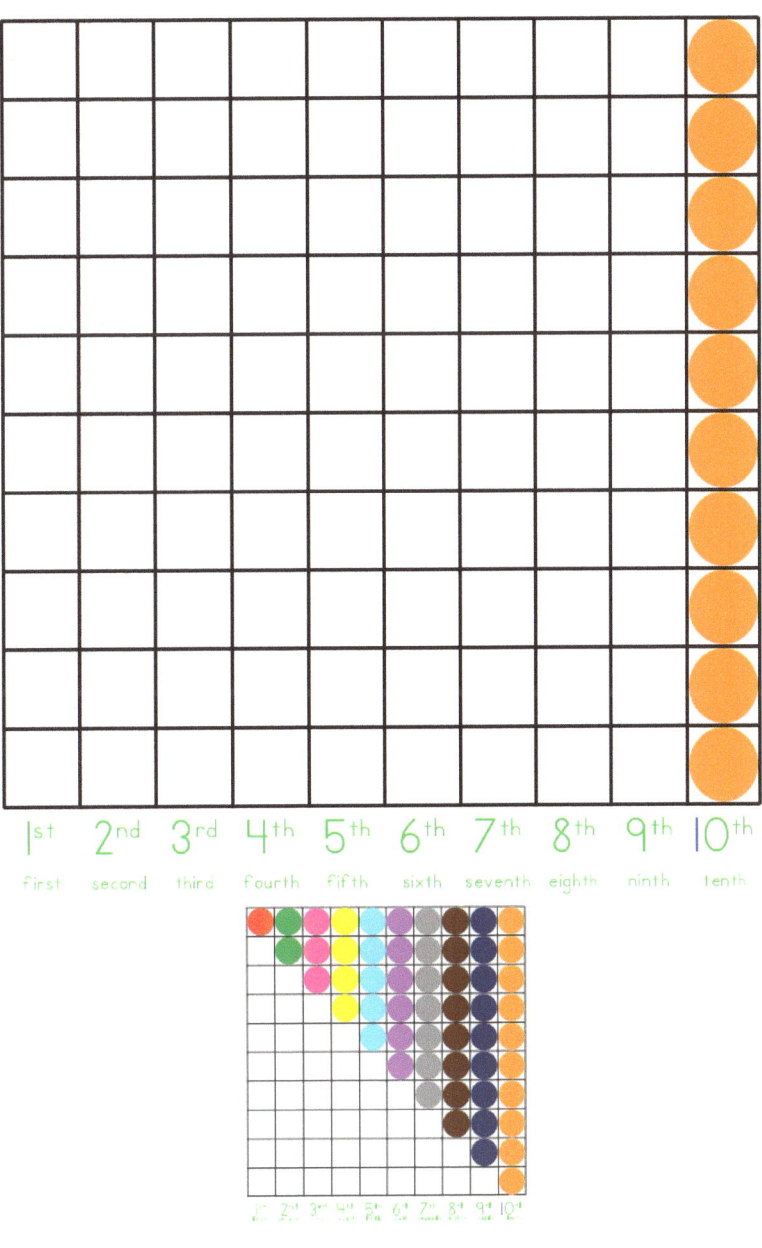

Exercise 3-6

Fill in the open spaces with the missing colored bead bars (as <u>sequential numeric positions</u>) in their vertical arrangement <u>and</u> write their <u>symbolic and verbal representations</u>.

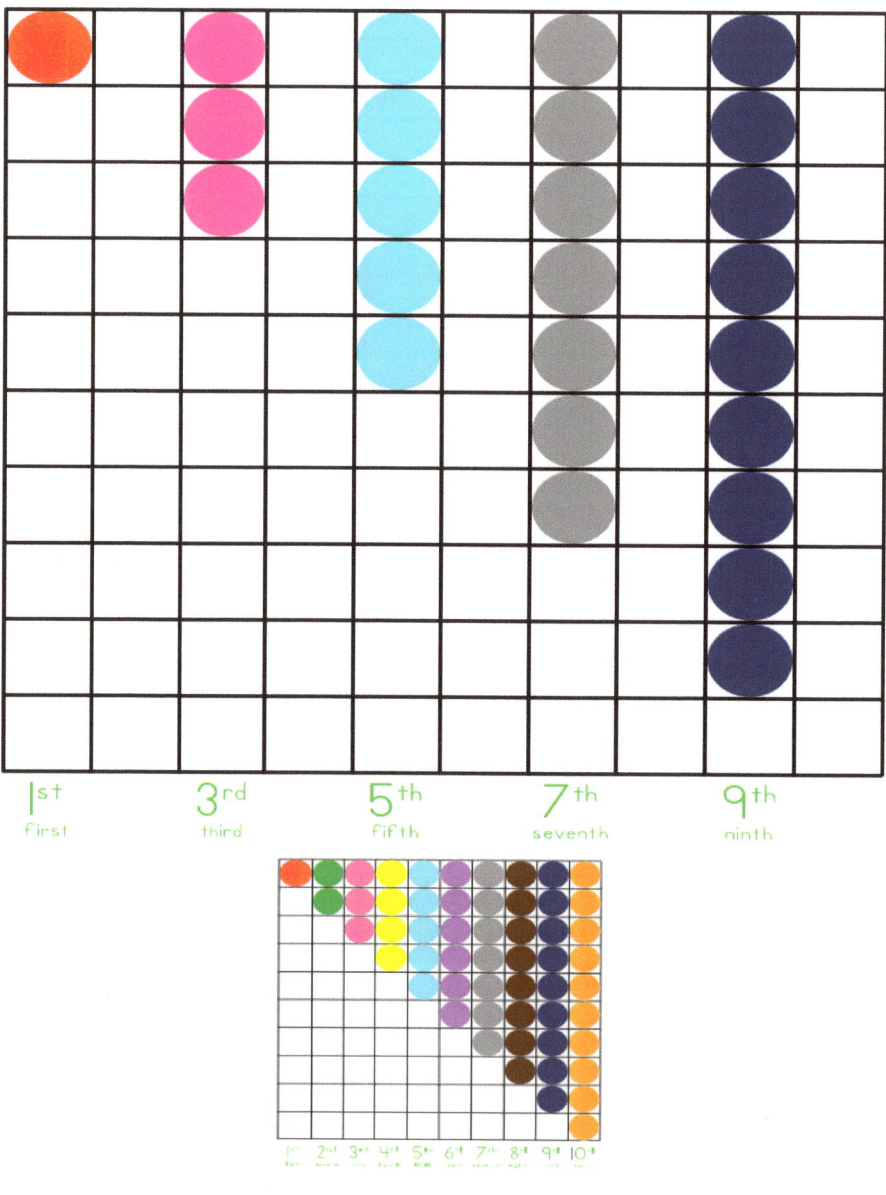

Exercise 3-7

Fill in the open spaces with the missing colored bead bars (as <u>sequential numeric positions</u>) in their vertical arrangement <u>and</u> write their <u>symbolic and verbal representations</u>.

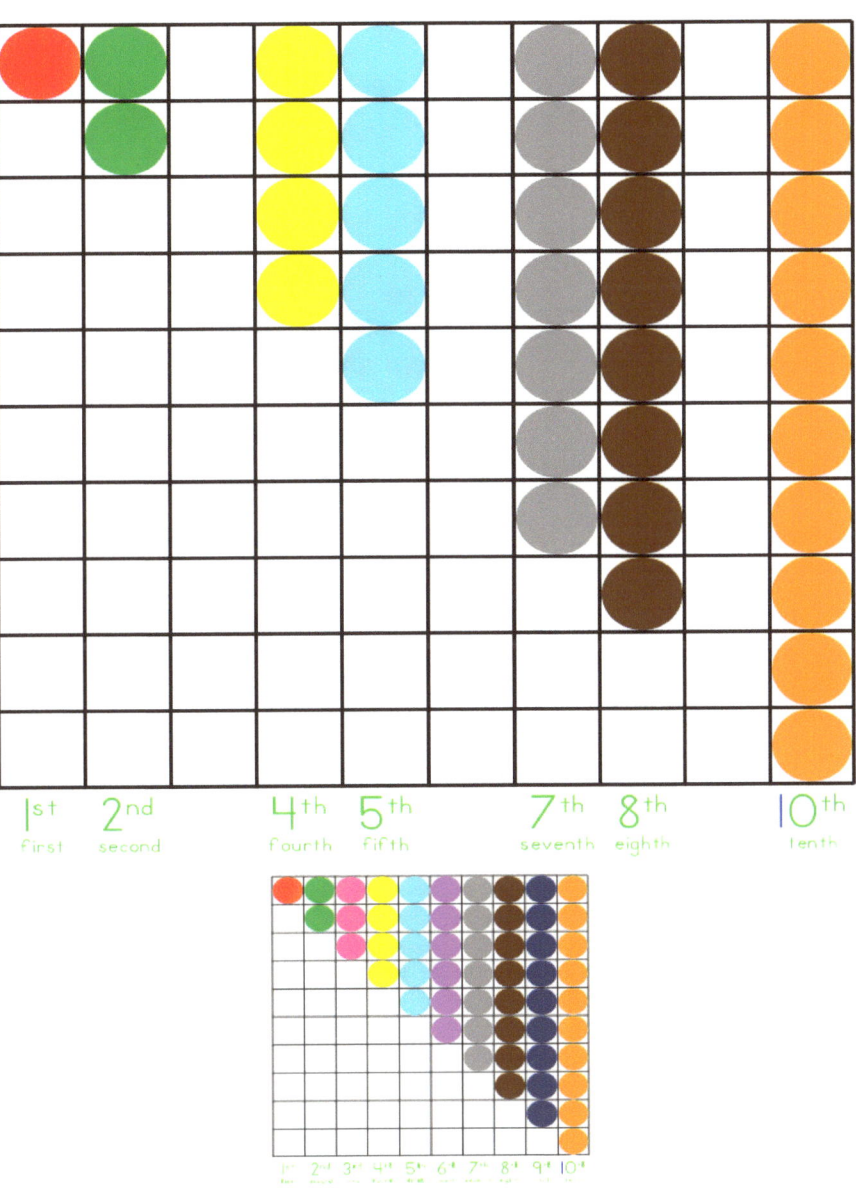

Exercise 3-8

Fill in the open spaces with the missing colored bead bars (as <u>sequential numeric positions</u>) in their vertical arrangement <u>and</u> write their <u>symbolic and verbal representations</u>.

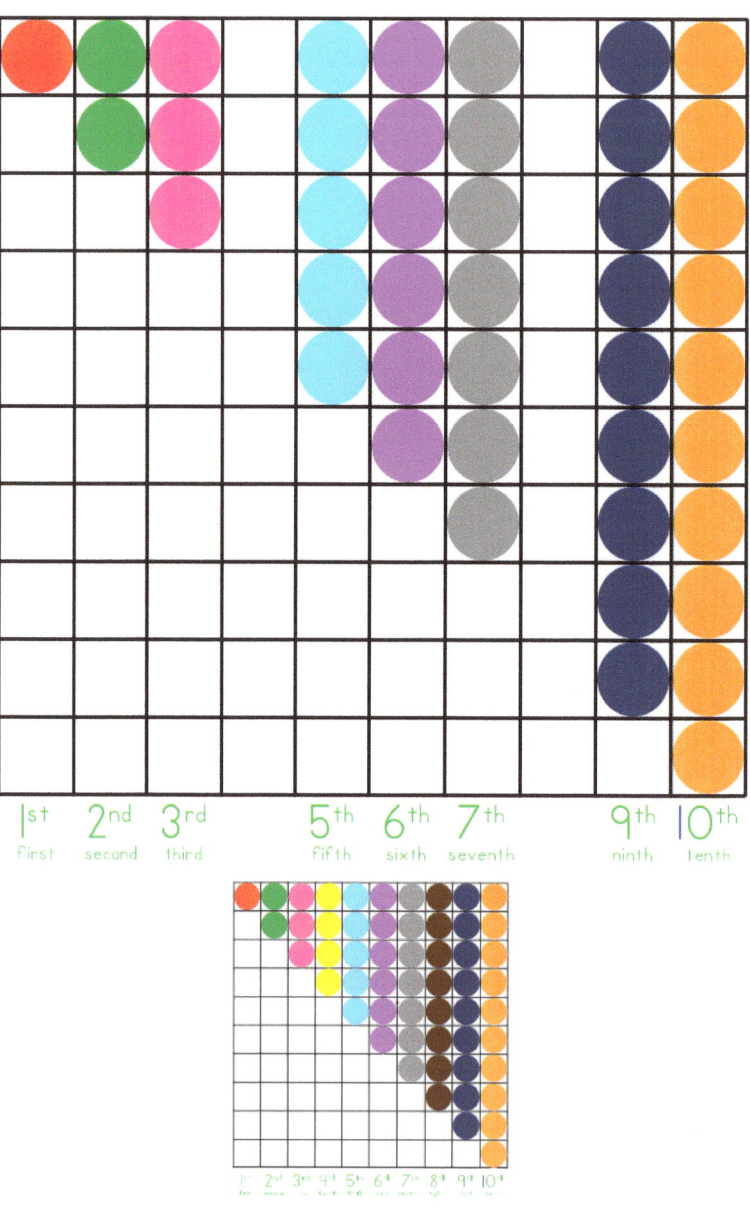

Exercise 3-9

Fill in the open spaces with the missing colored bead bars (as <u>sequential numeric positions</u>) in their vertical arrangement <u>and</u> write their <u>symbolic and verbal representations</u>.

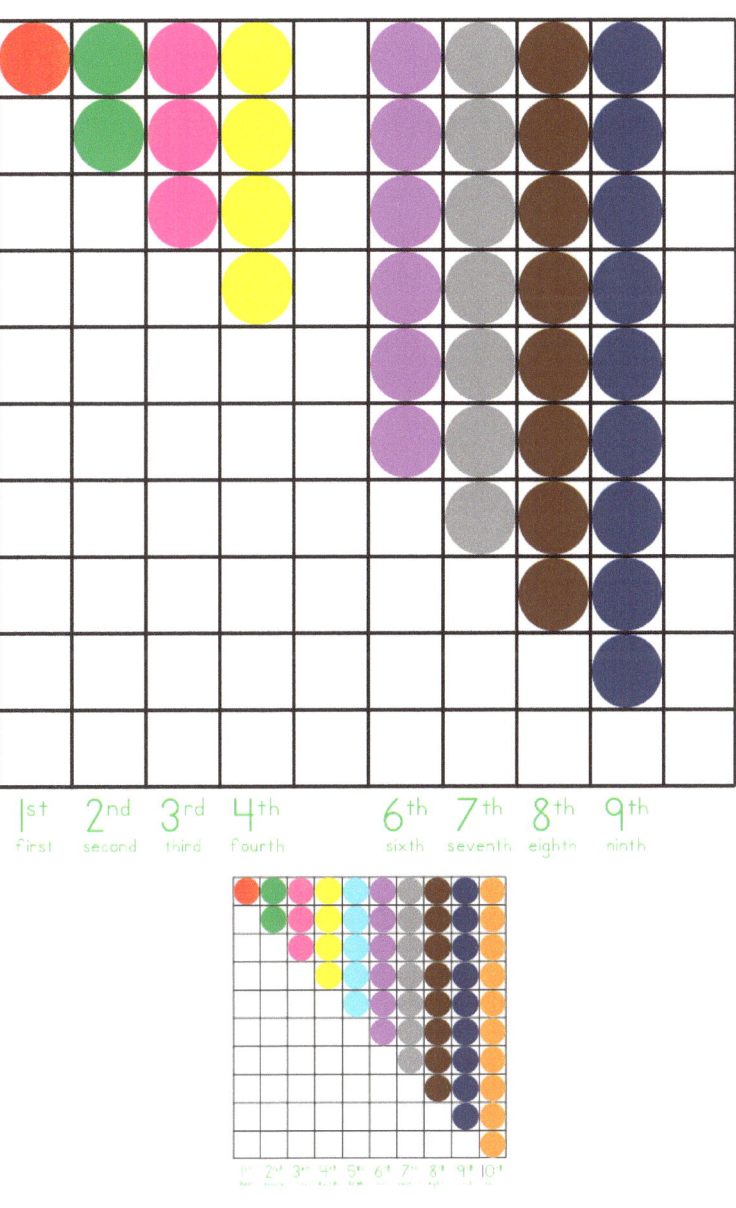

Exercise 3-10

Fill in the open spaces with the missing colored bead bars (as <u>sequential numeric positions</u>) in their vertical arrangement <u>and</u> write their <u>symbolic and verbal representations</u> from 1 – 9.

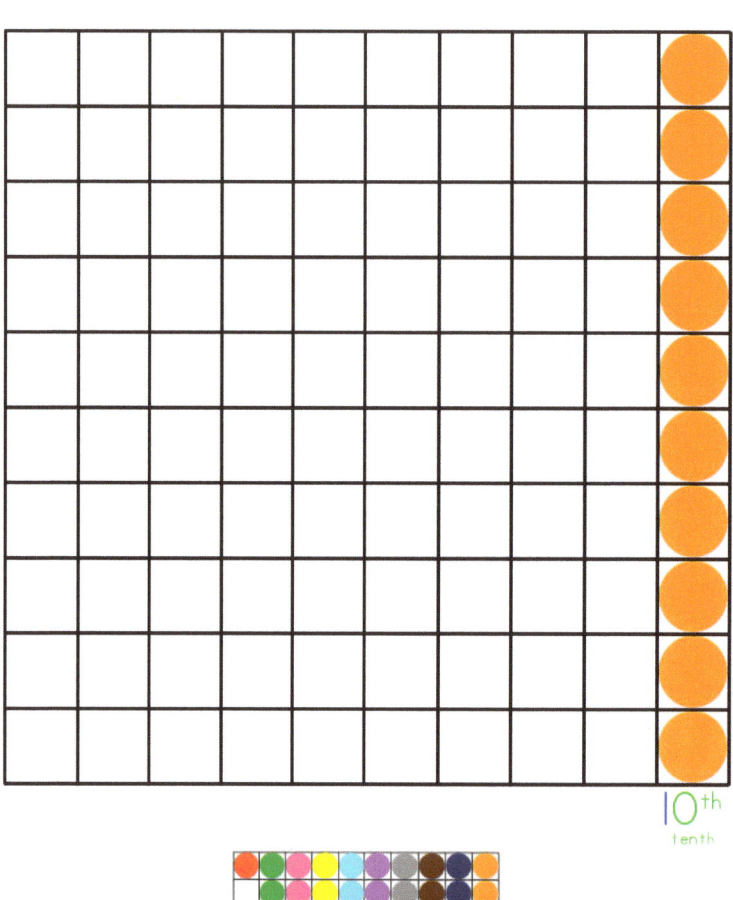

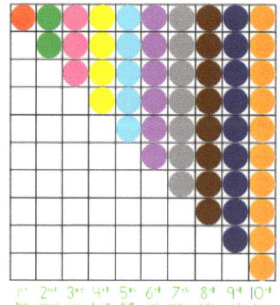

A Vision for Intellectual Wellness
In tribute to Dionicious and Mildred Gunawardena

Section 4

<u>Ordinal</u> Numbers in Horizontal Order

Presentation Chart 4

<u>Ordinal</u> Numbers in Horizontal Order

Illustrated below are the <u>quantities</u> and their associated <u>sequential numeric positions</u> as <u>ordinal numbers</u>. (Note the vocabulary underlined.)

- Observe the quantities arranged along a horizontal sequence.
- Observe the numerals associated with the quantities.
- Observe the particular color assigned to each quantity.

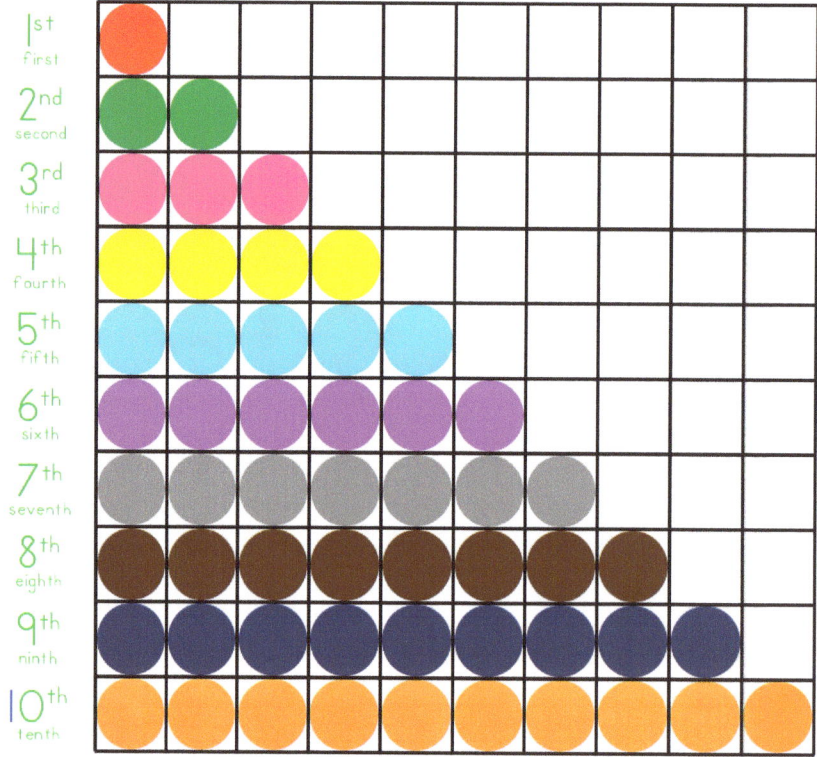

Exercise 4-1

Fill in the open spaces with the missing colored bead bars (as <u>sequential numeric positions</u>) in their horizontal arrangement.

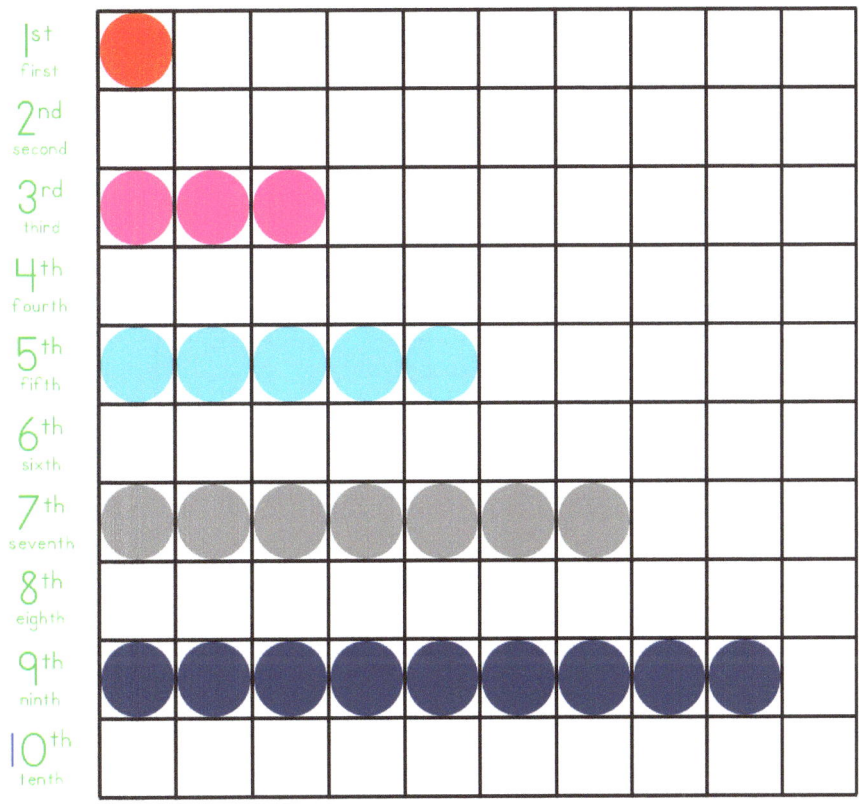

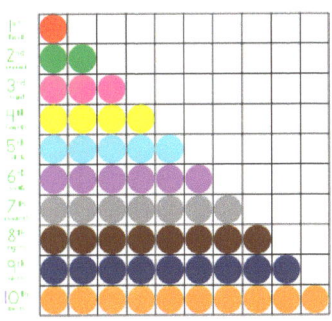

Exercise 4-2

Fill in the open spaces with the missing colored bead bars (as <u>sequential numeric positions</u>) in their horizontal arrangement.

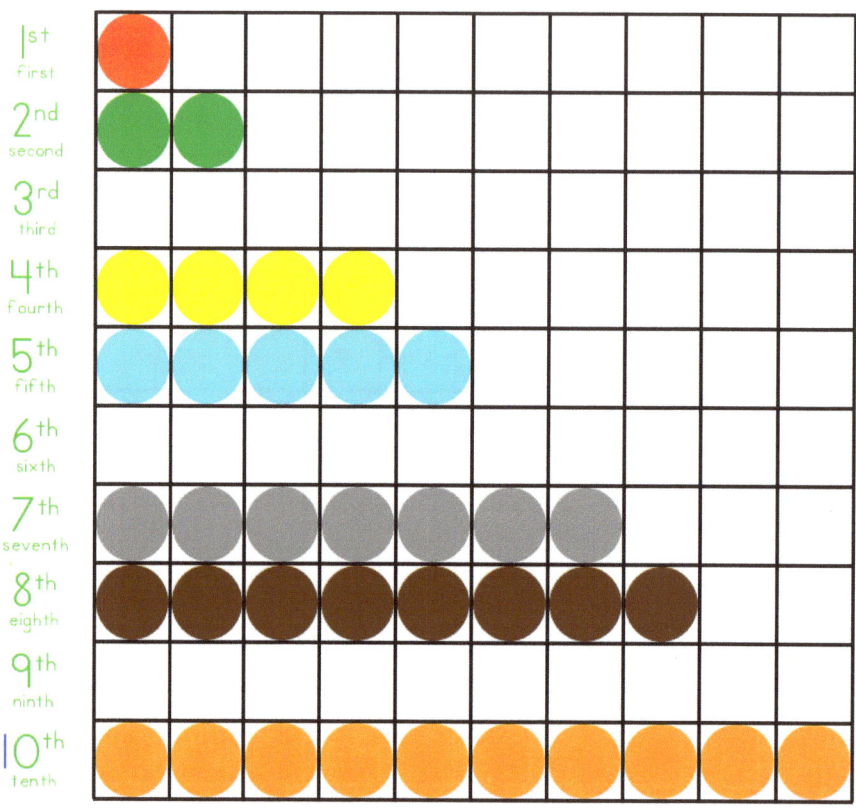

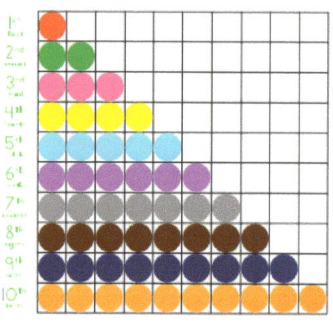

Exercise 4-3

Fill in the open spaces with the missing colored bead bars (as <u>sequential numeric positions</u>) in their horizontal arrangement.

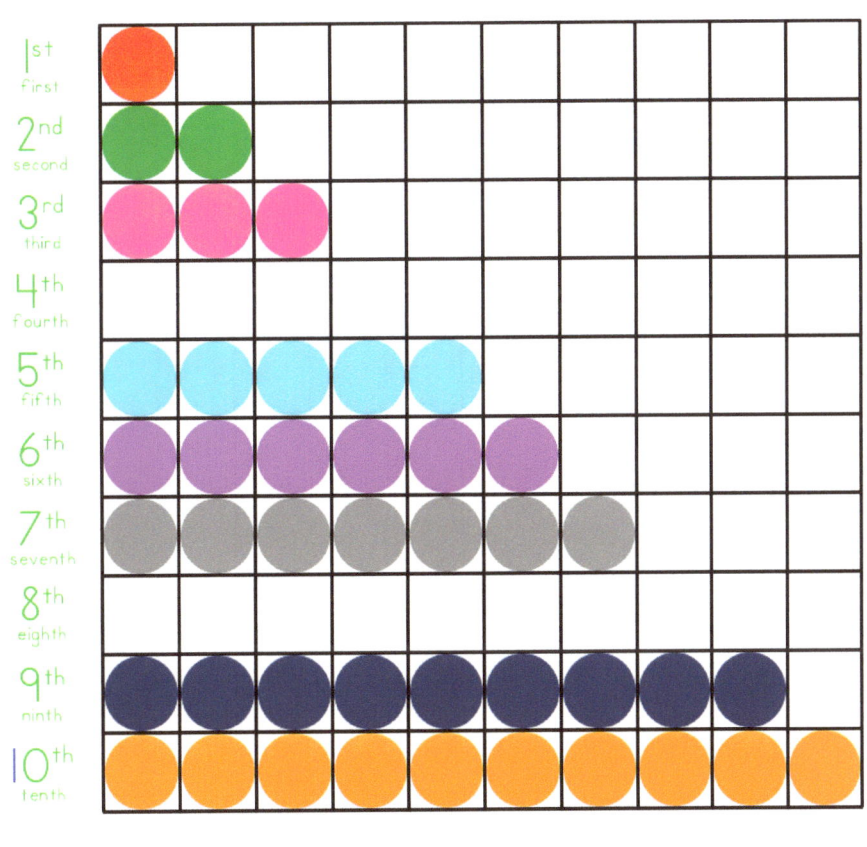

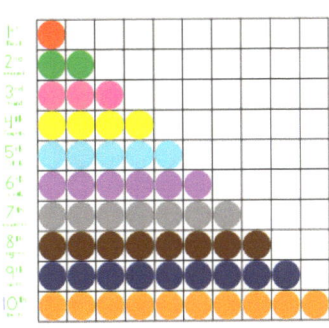

Exercise 4-4

Fill in the open spaces with the missing colored bead bars (as <u>sequential numeric positions</u>) in their horizontal arrangement.

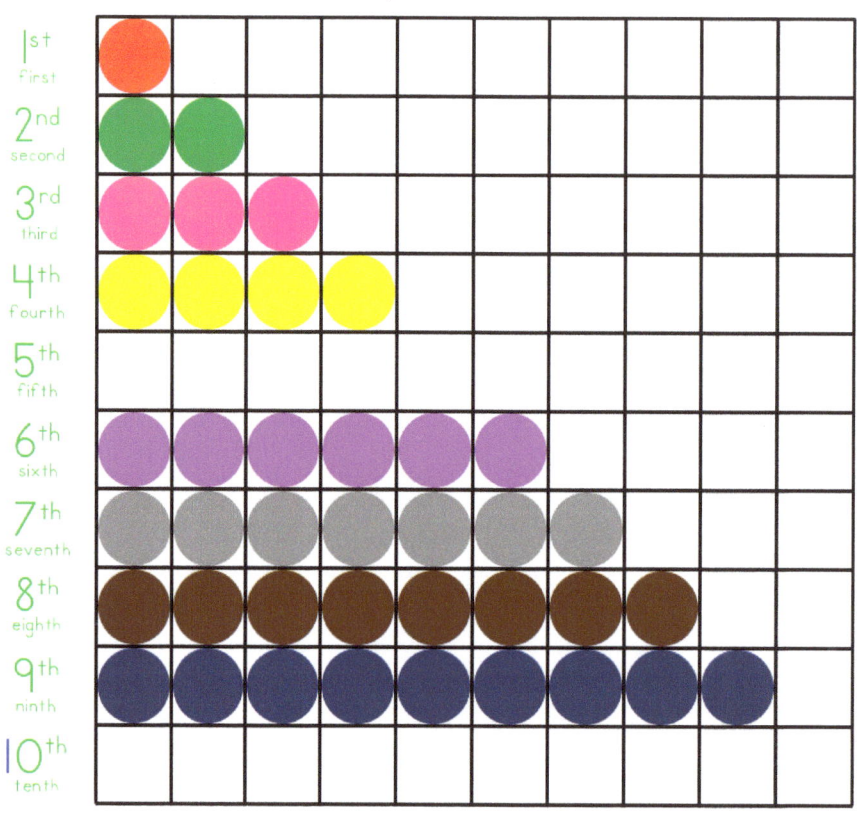

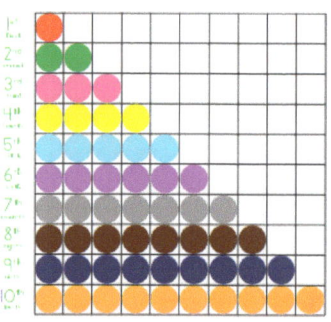

A Vision for Intellectual Wellness
In tribute to Dionicious and Mildred Gunawardena

Exercise 4-5

Fill in the open spaces with the missing colored bead bars (as <u>sequential numeric positions</u>) 1 – 9 in their horizontal arrangement, from top to bottom.

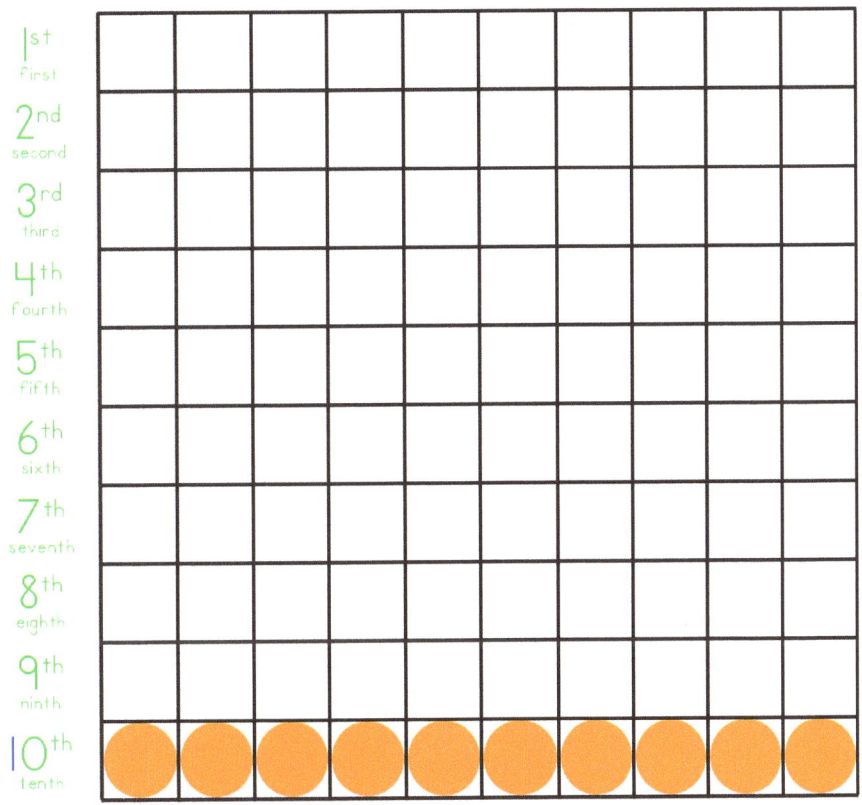

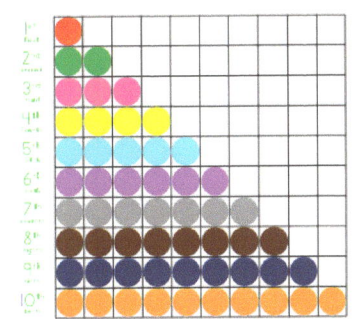

Exercise 4-6

Fill in the open spaces with the missing colored bead bars (as <u>sequential numeric positions</u>) in their horizontal arrangement <u>and</u> write their <u>symbolic and verbal representations</u>.

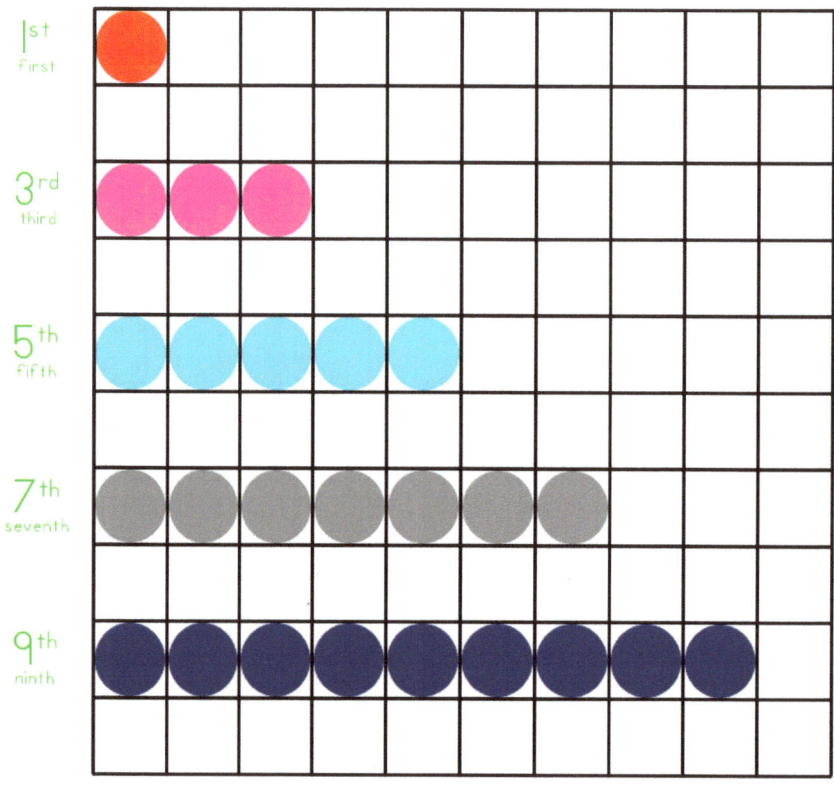

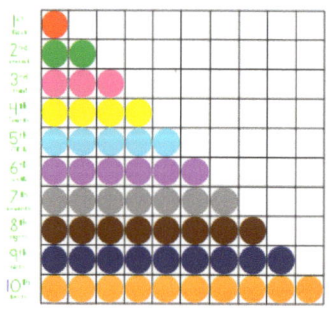

A Vision for Intellectual Wellness
In tribute to Dionicious and Mildred Gunawardena

Exercise 4-7

Fill in the open spaces with the missing colored bead bars (as <u>sequential numeric positions</u>) in their horizontal arrangement <u>and</u> write their <u>symbolic and verbal representations</u>.

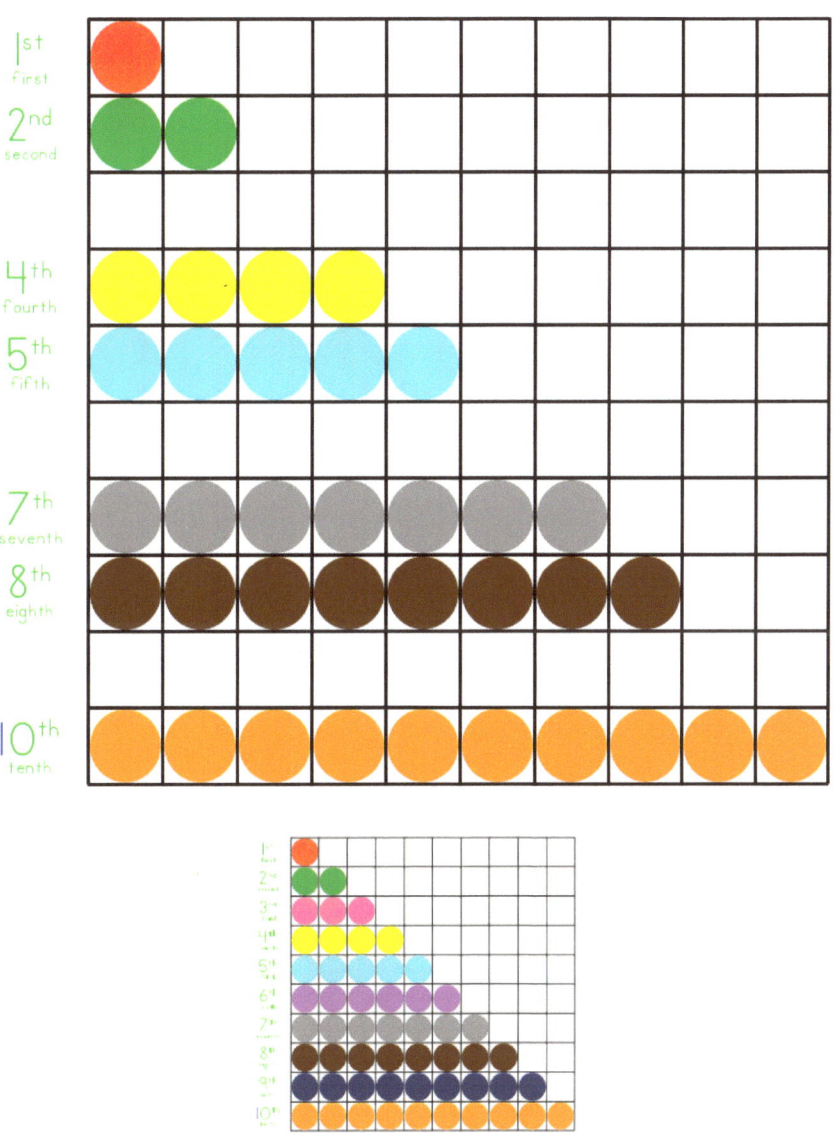

Exercise 4-8

Fill in the open spaces with the missing colored bead bars (as <u>sequential numeric positions</u>) in their horizontal arrangement <u>and</u> write their <u>symbolic and verbal representations</u>.

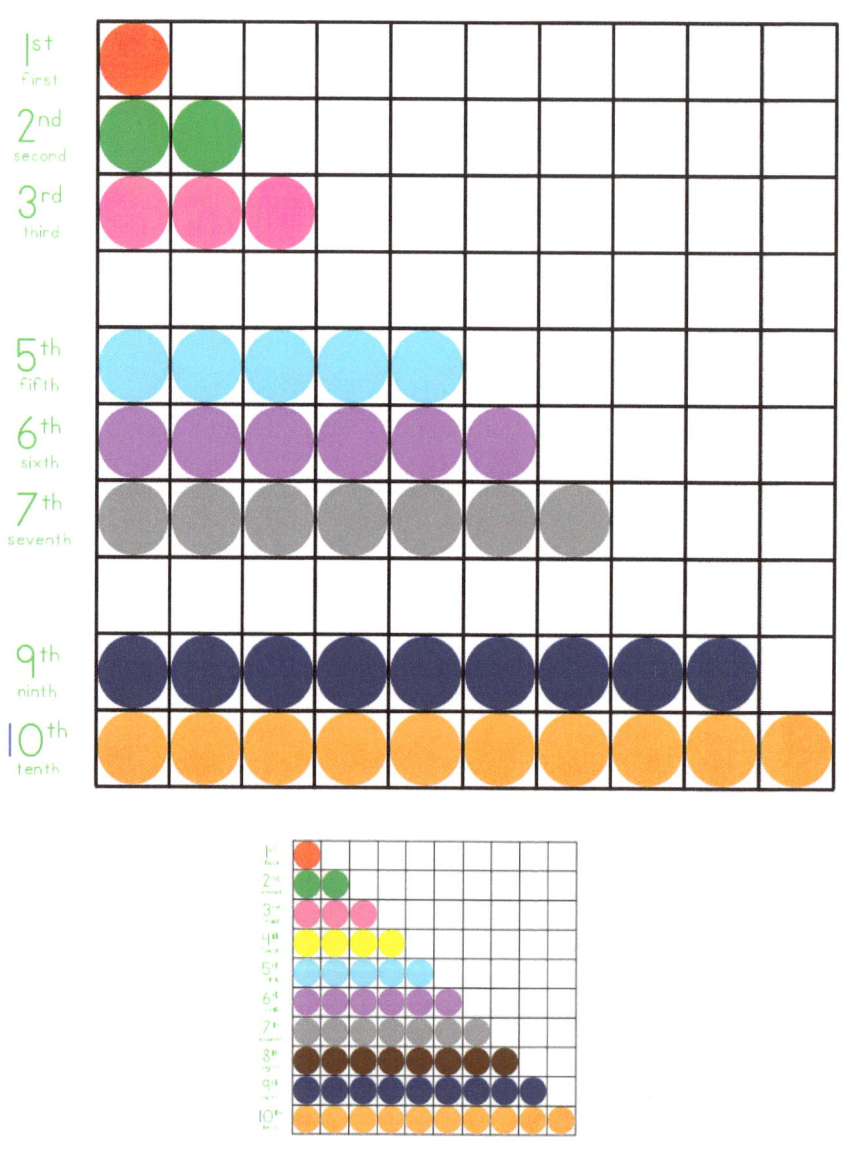

Exercise 4-9

Fill in the open spaces with the missing colored bead bars (as <u>sequential numeric positions</u>) in their horizontal arrangement <u>and</u> write their <u>symbolic and verbal representations</u>.

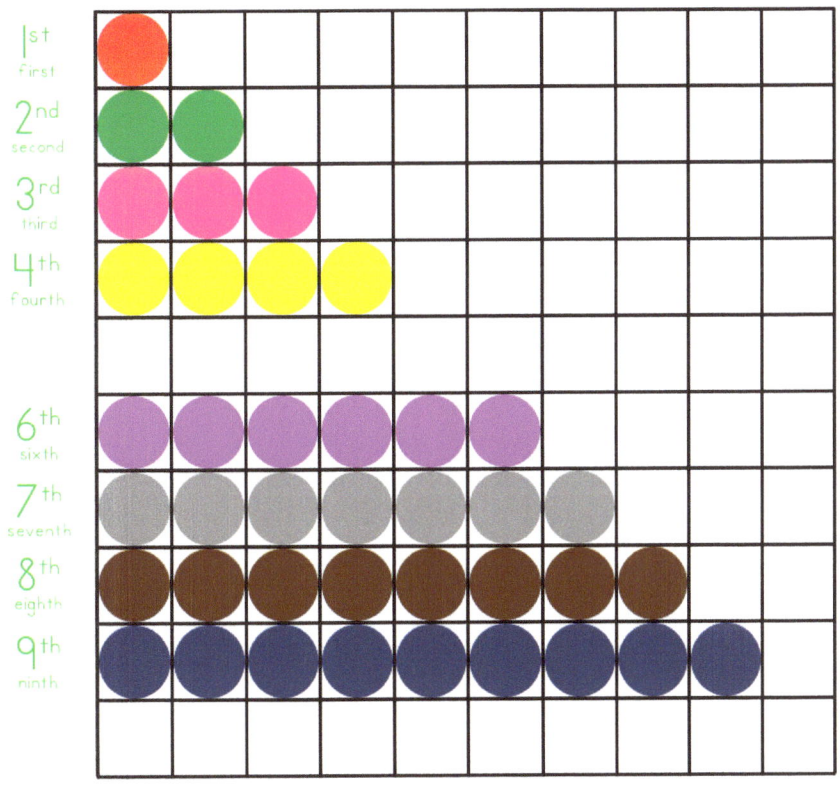

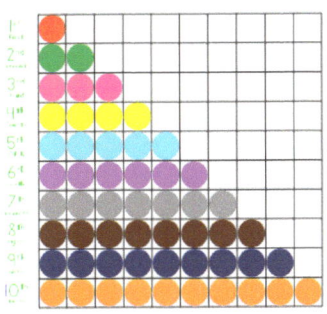

Exercise 4-10

Fill in the open spaces with the missing colored bead bars (as <u>sequential numeric positions</u>) in their horizontal arrangement <u>and</u> write their <u>symbolic and verbal representations</u> from 1 – 9.

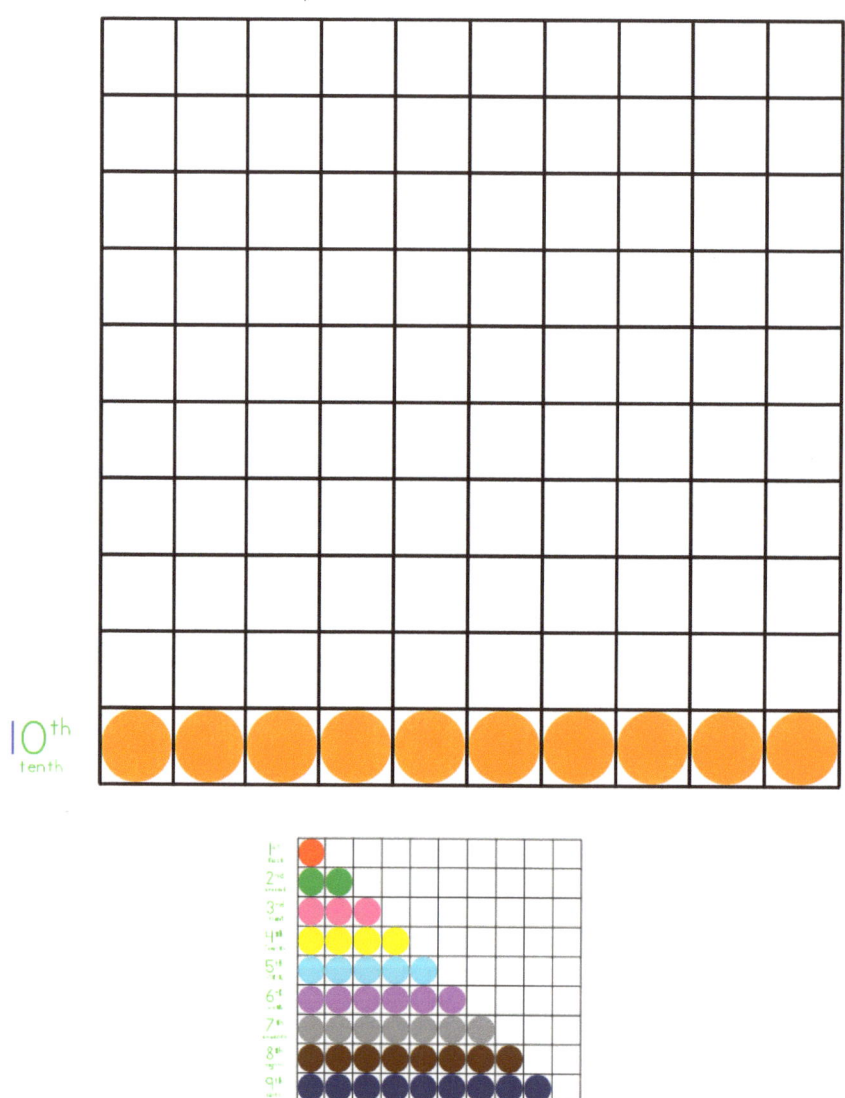

Review Exercise A – <u>Ordinal</u> Numbers
1-4 Vertical

Fill in the open spaces with the missing colored bead bars (or <u>quantities</u>) in their vertical arrangement <u>and</u> write their <u>numerical symbols</u>.

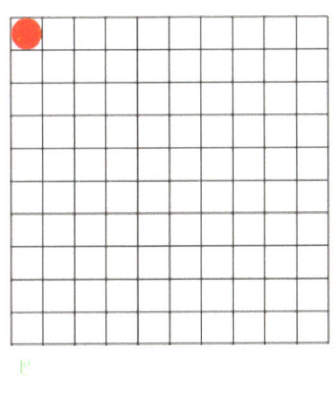

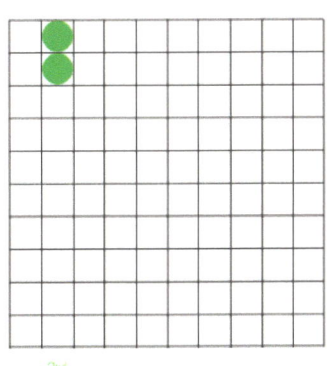

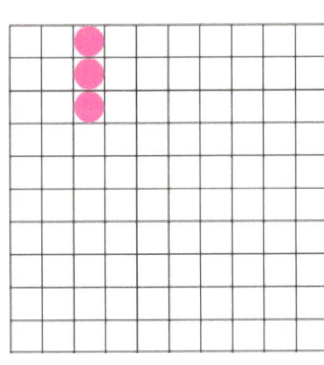

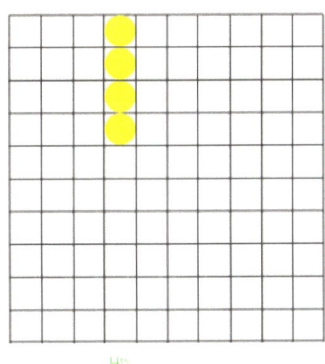

Review Exercise A – __Ordinal__ Numbers
5-8 Vertical

Fill in the open spaces with the missing colored bead bars (or __quantities__) in their vertical arrangement __and__ write their __numerical symbols__.

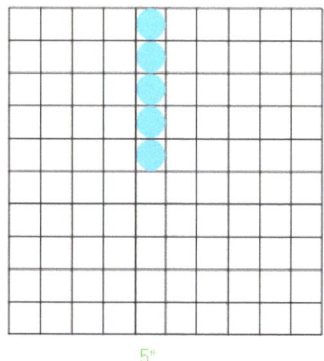

5th

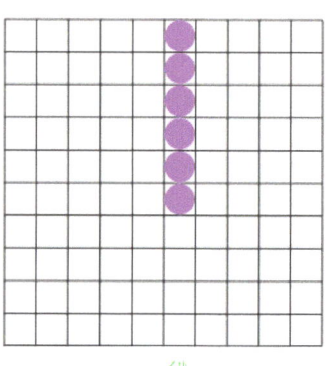

6TH

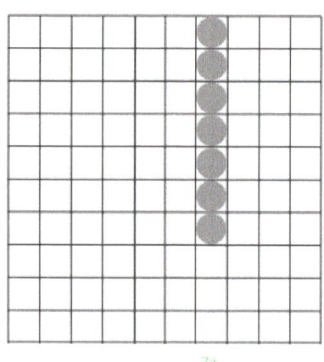

7⁺

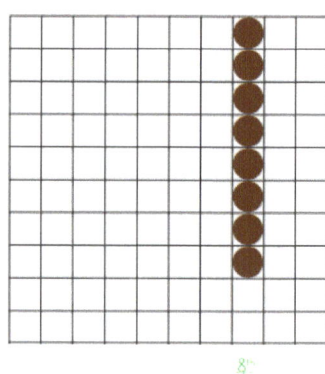

8th

A Vision for Intellectual Wellness
In tribute to Dionicious and Mildred Gunawardena

Review Exercise A – <u>Ordinal</u> Numbers
9-10 Vertical

Fill in the open spaces with the missing colored bead bars (or <u>quantities</u>) in their vertical arrangement <u>and</u> write their <u>numerical symbols</u>.

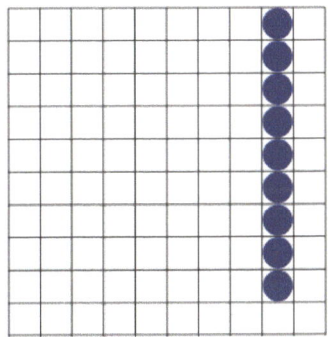

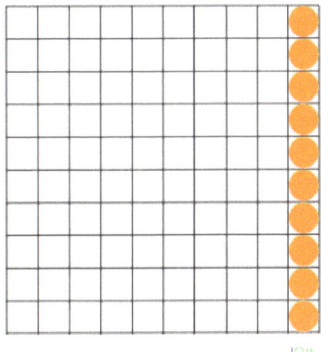

Review Exercise B – <u>Ordinal</u> Numbers
1-4 Horizontal

Fill in the open spaces with the missing colored bead bars (or <u>quantities</u>) in their horizontal arrangement <u>and</u> write their <u>numerical symbols</u>.

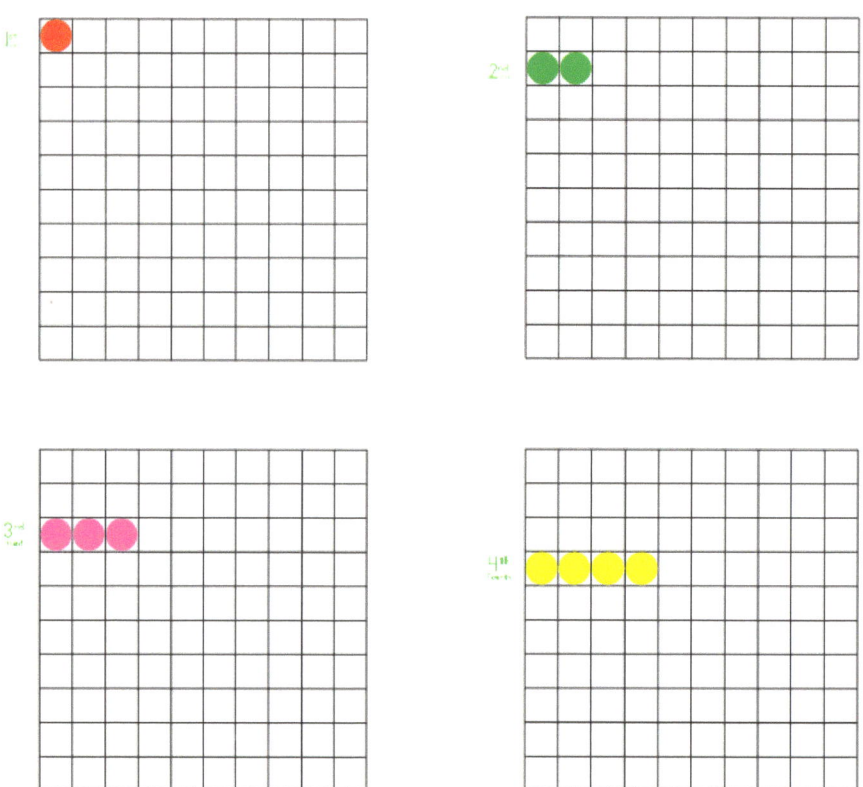

A Vision for Intellectual Wellness
In tribute to Dionicious and Mildred Gunawardena

Review Exercise B – <u>Ordinal</u> Numbers
5-8 Horizontal

Fill in the open spaces with the missing colored bead bars (or <u>quantities</u>) in their horizontal arrangement <u>and</u> write their <u>numerical symbols</u>.

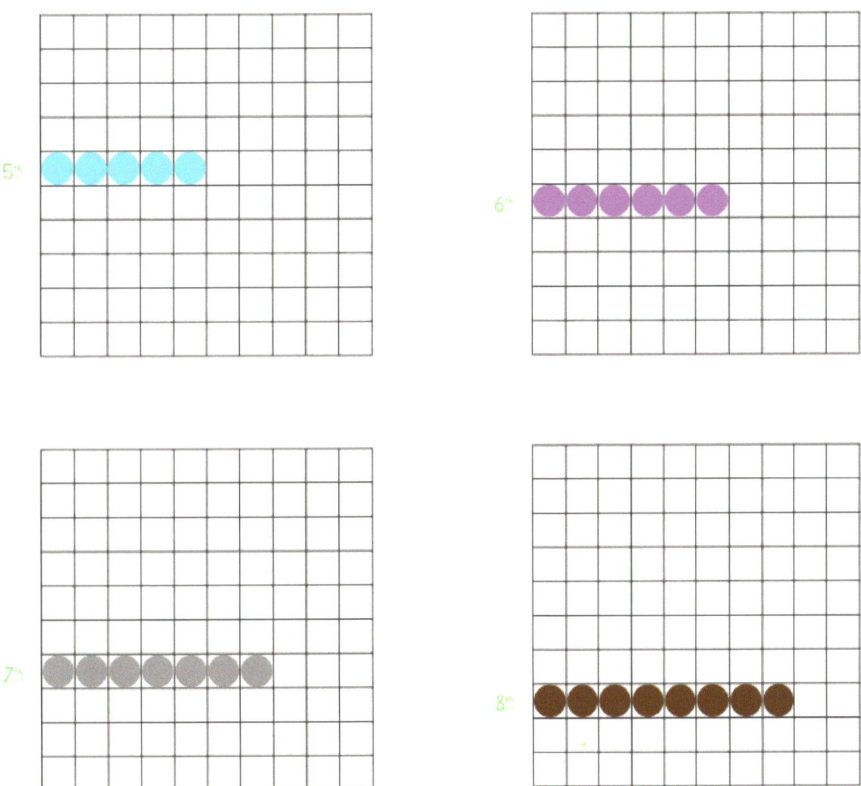

Review Exercise B – <u>Ordinal</u> Numbers
9-10 Horizontal

Fill in the open spaces with the missing colored bead bars (or <u>quantities</u>) in their horizontal arrangement <u>and</u> write their <u>numerical symbols</u>.

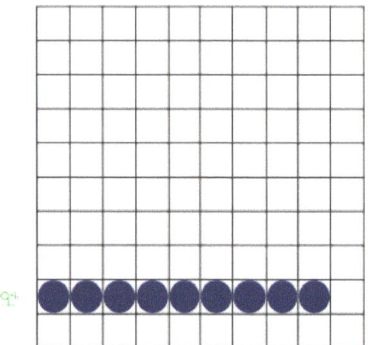

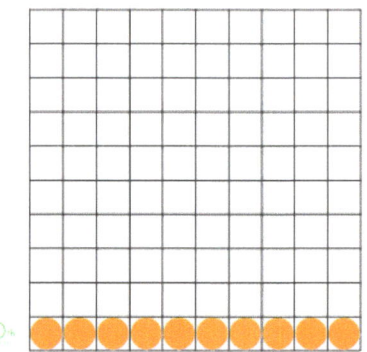

A Vision for Intellectual Wellness
In tribute to Dionicious and Mildred Gunawardena

Review Exercise C – <u>Ordinal</u> Numbers
1-10 Vertical and Horizontal Positions Align to Form Squares

Observe the positional pattern of colored bead bars (or <u>quantities</u>) arranged in their simultaneously vertical and horizontal sequence.

EXAMPLE:

Review Exercise C – __Ordinal__ Numbers
1-4 Vertical and Horizontal Positions Align to Form Squares

Fill in the open spaces with the missing colored bead bars (or __quantities__) in their vertical and horizontal arrangement __and__ write their __numerical symbols__.

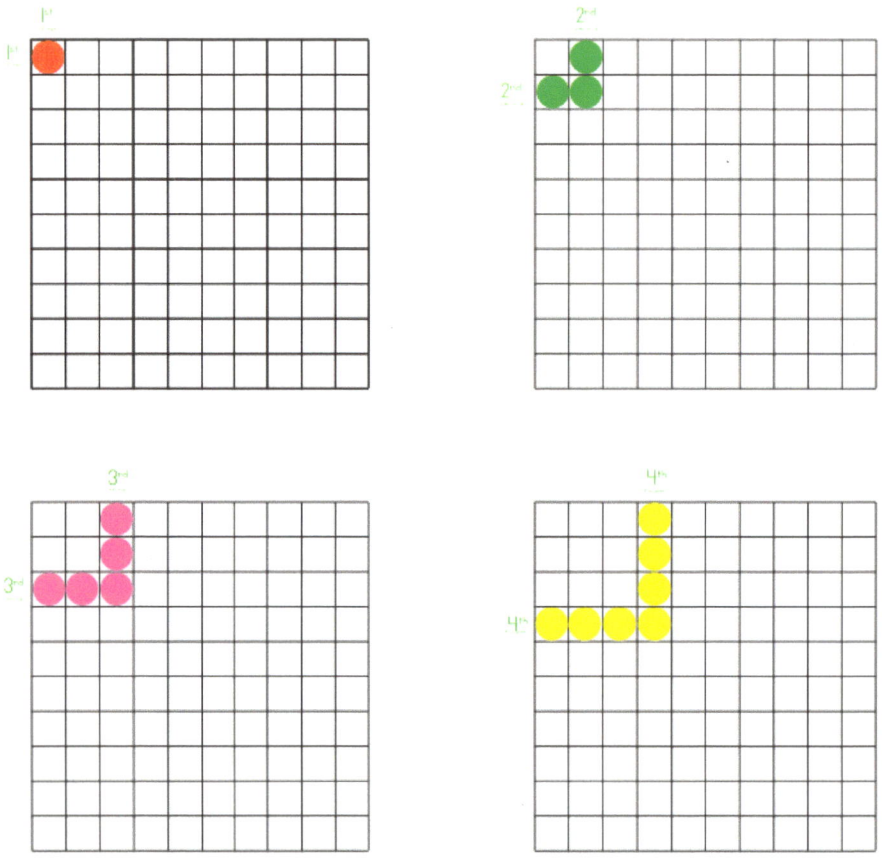

Review Exercise C – <u>Ordinal</u> Numbers
5-8 Vertical and Horizontal Positions Align to Form Squares

Fill in the open spaces with the missing colored bead bars (or <u>quantities</u>) in their vertical and horizontal arrangement <u>and</u> write their <u>numerical symbols</u>.

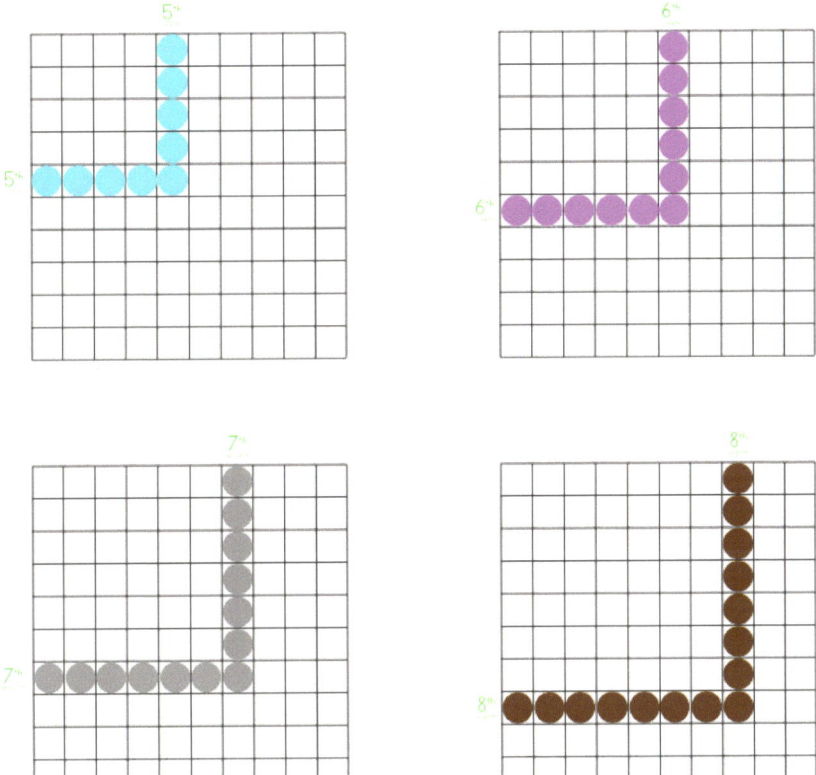

Review Exercise C – <u>Ordinal</u> Numbers
9-10 Vertical and Horizontal Positions Align to Form Squares

Fill in the open spaces with the missing colored bead bars (or <u>quantities</u>) in their vertical and horizontal arrangement <u>and</u> write their <u>numerical symbols</u>.

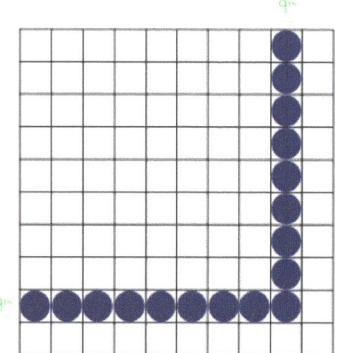
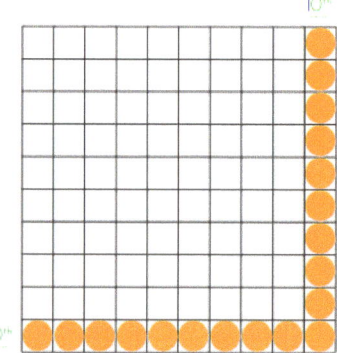

A Vision for Intellectual Wellness
In tribute to Dionicious and Mildred Gunawardena

Review Exercise D – <u>Ordinal</u> Numbers
1-10 Vertical and Horizontal Positions Align to Form Squares

Fill in the color defined quantities to match their associated numerals as shown in their vertical and horizontal sequence.

	1st first	2nd second	3rd third	4th fourth	5th fifth	6th sixth	7th seventh	8th eighth	9th ninth	10th tenth
1st first										
2nd second										
3rd third										
4th fourth										
5th fifth										
6th sixth										
7th seventh										
8th eighth										
9th ninth										
10th tenth										

Review Exercise E – <u>Ordinal</u> Numbers
1-10 Vertical and Horizontal Positions Align to Form Squares

Fill in the color defined quantities in their vertical and horizontal sequence along with their associated numerical positions.

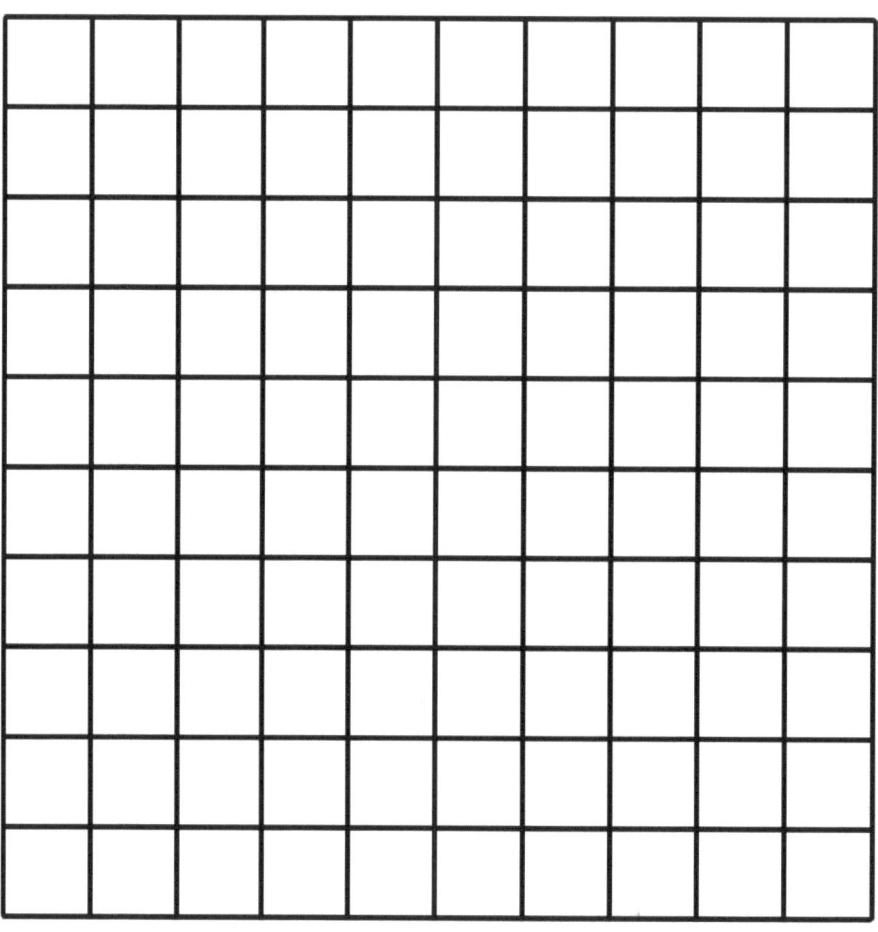

An original from The Precious Jewels of Mrs. "G"
Basic Number Concepts 1-9

Practice Template

A Vision for Intellectual Wellness
In tribute to Dionicious and Mildred Gunawardena

This is in appreciation of engaged-partnerships dedicated towards the cultivation and preservation of the active brain.

Thank you, for your sincere participation. You have been given a precise introduction towards the attainment of perfection and exactitude along patterns of numeration, as the most basic threshold to the stairway of mathematical advancement and the preservation of the intellect.

We do hope you derived a benefit from this presentation. Consistent practice, indeed, by means of the exercises of repetition—presented each time, with a new twist—as shown herein, will prepare you to further absorb and assimilate the dynamics of the numerical concepts—indispensable for the cultivation of deeper understanding in regard to the many tiers of intellectual mathematical evolution.

In order to reap the fullest benefit from the books in the series **The Active Brain**, it is important that you do not skip, out of sequence, in your future acquisitions of volumes in this collection. It is highly recommended that you practice the exercises in each book along the sequence until near perfection, before advancing to the next book.

A Vision for Intellectual Wellness
In tribute to Dionicious and Mildred Gunawardena

Please note

The author does not take responsibility for any misconceptions on the part of the users—of this series of books—who may not adhere to the principles of presentation, as aforementioned.

A Vision for Intellectual Wellness
In tribute to Dionicious and Mildred Gunawardena

Vision, Design, and Presentation by
Nelunika Gunawardena Rajapakse

Gracious Technical Support By
Debra Murray

A Vision for Intellectual Wellness
In tribute to Dionicious and Mildred Gunawardena

About the Author

Nelunika Gunawardena Rajapakse is a proud citizen of her world, long in residence over four decades in the United States. She is the only daughter, besides two precious sons, of late Dionicious and Mildred Gunawardena who arrived in the USA, by invitation. The latter was trained directly under Dr. Maria Montessori in 1943, and it was her pride and joy to share the original vision of Dr. Montessori amidst a growing interest, then, in the United States, to understand the Montessori perspective in its authentic sense.

Nelunika, needless to say, has been a product of this unmatched original vision from the very inception of her life, to this day. Mentored all along by her resolutely dedicated Montessori pedagogue mother, and having worked in partnership with her for three decades, Nelunika is a proud beneficiary of her own experientially rich, intellectual legacy.

She boasts a rich history of authentic Montessori in practice for over four decades, having been the head directress, administrator, and chief operating officer of Bainbridge~Solon Montessori School (1970-2012).

The author lives her renewed and pioneering vision presently along her steadfast mission to 'Raise the Position of Today's Child and the World' while sharing her 'Vision for Intellectual Wellness for all Ages'. As such, she disseminates her wisdom and expertise through the creation of invaluable time-stamps of her life-experience.

Additionally, Mrs. Nelunika Rajapakse is a devoted professional trainer, and a mentor for parents, engaged in sharing her expertise by way of lectures, seminars, consultations, and continuous guidance to organizations, schools, individuals, and diverse social groups.

A Vision for Intellectual Wellness
In tribute to Dionicious and Mildred Gunawardena

More information may be sought by visiting the following websites:

Where to find soft copy editions of books in print:

http://www.amazon.com/author/nelunikarajapakse
http://www.createspace.com/3694500 (Compassionate, Engaged Parenting as a Help to Life)
http://www.createspace.com/3888998 (Parenting as a Help to Life– revised edition 2012)
http://www.createspace.com/900002015 (Roots of Moral Sensitivity and the Emerging Tree of Life)
http://www.createspace.com/3992465 (The Active Brain Book 1)
http://www.createspace.com/4086052 (The Active Brain Book 2, Volume 1)
http://www.createspace.com/4368065 (The Active Brain Book 2, Volume 2)

Where to find hard copy editions of books in print:

http://www.shop.greetingsofdharma.com

Where to find Kindle editions of books in print:

http://www.amazon.com

Where to find additional e-books and greetings authored by Nelunika Gunawardena ~ Rajapakse:

http://www.shop.greetingsofdharma.com

To find updates related to books authored by Nelunika Gunawardena Rajapakse and other pertinent information on parenting, please visit:

http://www.twitter.com/Nelunika2010
http://www.linkedin.com/NelunikaGunaRaja
http://www.parentingasahelptolife.blogspot.com
http://www.facebook.com/Nelunika2010

All Associated Websites:

http://www.parentingasahelptolife.org
http://www.greetingsofdharma.com
http://www.shop.greetingsofdharma.com
http://www.amazon.com/author/nelunikarajapakse
http://www.bainbridgemontessori.org

www.ingramcontent.com/pod-product-compliance
Lightning Source LLC
Chambersburg PA
CBHW051020180526
45172CB00002B/419